T0181449

An Integrated Modelling Approach to Design Cost-Effective AES for Agricultural Soil Erosion and Water Pollution

Zhengzheng Hao

An Integrated Modelling Approach to Design Cost-Effective AES for Agricultural Soil Erosion and Water Pollution

An Application of a Real Watershed in China

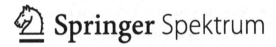

Zhengzheng Hao
Cottbus, Germany

A Novel Integrated Modelling Approach to Design Cost-effective Agri-Environment Schemes to Prevent Soil Erosion and Water Pollution from Cropland — A Case Study of Baishahe Watershed in Shanxi Province, China.
A thesis approved by the Faculty of Environment and Natural Sciences at the Brandenburg University of Technology in Cottbus-Senftenberg in partial fulfilment of the requirement for the award of the academic degree of Doctor of Philosophy (Ph.D.) in Environmental Sciences.

ISBN 978-3-658-41339-2   ISBN 978-3-658-41340-8   (eBook)
https://doi.org/10.1007/978-3-658-41340-8

This Springer Spektrum imprint is published by the registered company Springer Fachmedien Wiesbaden GmbH, part of Springer Nature.
The registered company address is: Abraham-Lincoln-Str. 46, 65189 Wiesbaden, Germany

I would like to dedicate my work to my big family for their selfless supports in any forms, to my PhD study supervisor Prof. Frank Wätzold for his mindful and patient guidance, to my dear friends for their spiritual care with pure hearts, and to my own striving youth.

Acknowledgements

First and foremost, I would like to thank my supervisors, Prof. Frank Wätzold and Prof. Frank Molkenthin, and Dr. Astrid Sturm. All of them have contributed so much to turn an unconfident poor learner into the one prepared to get the PhD degree and plan the career in academia. Prof. Frank Molkenthin gave his precious advice and comments for me to fulfill and complete my interdisciplinary work regarding the part of eco-hydrological modelling. Dr. Astrid Sturm, a very special nice lady with her amazing hair, has impressed me so many times with her clear logic, quick understanding of my bad expression, patience and enthusiasm to me. Prof. Frank Wätzold, the savior I met in life with the very best of my luck, is a great mentor for guiding both my research and life. He has high professional knowledge and imparts it with enormous amounts of patience in the nicest way. At the same time, his diligence, integrity, kindness and positive energy have influenced me a lot. He and Astrid have helped and encouraged me so much for getting confidence during my study.

Meanwhile, my heartfelt appreciation goes to other nice fellows in the Chair of Environmental Economics for their friendship, kindness and all kinds of support. They are Mrs. Regina Kirsche, Oliver Schöttker, Lutz Philip Hecker, Nonka Markova-Nenova, Charlotte Gerling, Mary Nthambi, Catrin Spring, Johanna Götter and others. Mrs. Regina Kirsche, the wonderful secretary in the Chair, I want to express my deep gratefulness to you for helping to manage all kinds of administrative processes for me and other helps and concerns in my personal life as a foreigner. Oliver Schöttker, thank you for your always earnest help for my research and life questions. Lutz Philip Hecker, your open-minded, easy-going personality and sense of humor just make me to ask for help without hesitation, but I never take it as granted, please accept my lots of thanks. Mary Nthambi, thank you for being my officemate and intimate friend, you will always be my life model as a great woman.

Further, I want to thank the PhD Program Environmental and Resource Management (ERM) of the Brandenburg University of Technology Cottbus-Senftenberg (BTU) and the involved people, namely Prof. Manfred Wanner, Dr. Birte Seffert, Prof. Michael Schmidt, Dr. Dmitry Palekhov, and Prof. Frank Wät-zold. I have improved so much with my presentation skills because of the training during this PhD ERM program. I appreciate very much the advice and comments from Prof. Manfred Wanner, the great coordination and management for us from Dr. Birte Seffert, and the very conscientious training for all perspectives of presentation from my supervisor Prof. Frank Wätzold. All I have learned from this would be my precious treasure in later career life.

Moreover, without the financial support I would never have a chance to start my PhD study on abroad. My sincere thanks go to the China Scholarship Council (CSC) for offering me a scholarship of four years for PhD study abroad. The same thanks go to the BTU Graduate Research School for offering me the short-term scholarship of six month for me to complete my PhD without huge financial pressure. At the same time, I acknowledge very much for the opportunities the BTU Graduate Research School offered me to get training with teaching and research skills under the teaching and research assistantships.

Last but not least, I wish to give my gratitude to the officers in relevant government sectors and the farmers in my study region for the cooperation and support during my data collection process. Also, I thank my big family for all your love, selfless support and all my met fellows on abroad for your companionship and care.

About the Book (English)

Cropland intensification is a key driver of soil erosion and water pollution from agricultural production in China and many other parts of the world. China's area accounts for 6.8% of the world, while its area of soil erosion accounts for 14.2% of that of the world. As the main source, sloping cropland accounts for 1/3 of the total national amount of soil erosion in China. On the other hand, China's agricultural pollution contributes about half of the total national water pollution load, and again cultivation practices on cropland are major contributors.

A frequent policy response for non-point source (NPS) environmental problems is to implement agri-environment schemes (AES). In AES, compensation payments are given to farmers for them to implement land use measures which are beneficial to environment. AES are more often practiced in developed countries than in developing countries. One of the key concerns for AES design is to optimize the cost-effectiveness. To achieve the aim, models that integrate economic and eco-hydrological knowledge have been developed, for assessing both the ecological effects and economic costs of agricultural land use measures. Although such integrated models are able to make recommendations on which land use measures should be implemented and where, the models typically adopt a planning perspective and do not take into account that farmers make voluntary decisions about whether to participate in an AES. Moreover, they ignore that an AES often consists of several measures among which farmers can choose and that this choice has to be considered in the design of an AES.

Based on this, this thesis presents a spatially explicit novel integrated modelling approach that addresses these shortcomings by combining several components. First is to identify appropriate candidate land use measures for the study region. Second is to quantify the ecological effects of land use measures regarding soil erosion and water pollution from nitrogen and phosphorus in spatially heterogeneous units through SWAT model simulation. Third is to assess the costs for land use measures in spatially heterogeneous units. Fourth is operating an optimization procedure that considers farmers' behavior and is able to design cost-effective AES to reduce soil erosion and water pollution (from N, P). Here cost-effectiveness means that under given budget for AES project the (weighted) reduction load of sediment from soil erosion and N and P from water pollution is minimized.

The relevance of the modelling procedure is demonstrated in the thesis by applying it to the Baishahe watershed, an area of approximately 56 km^2, in Shanxi province in China. The results show that the modelling approach is robust to design cost-effective AES, i.e. under given budget the maximum ecological effect could be attained with determined payments for measures. The developed modelling approach is generic and powerful for application in all kinds of agricultural watersheds with various sizes.

Keywords: Agri-environment schemes · Cost-effectiveness · Integrated modelling approach · Soil erosion · Water pollution

About the Book (Deutsch)

Die Intensivierung der Ackerlandnutzung ist eine Hauptursache für Bodenerosion und Wasserverschmutzung durch landwirtschaftliche Produktion in China und in vielen anderen Regionen der Welt. Während China einen Anteil von 6,8% an der weltweiten Landmasse hat, ist es für 14,2% der weltweiten Bodenerosion verantwortlich. Mit einem Anteil von 1/3 ist Ackerbau in Hanglagen eine Hauptursache für die Bodenerosion in China. Gleichzeitig geht über die Hälfte der nationalen Wasserschadstoffbelastung auf landwirtschaftliche Verschmutzungen zurück, wobei Anbaumethoden auf Ackerland erneut hauptursächlich sind.

Eine häufige Reaktion der Politik auf Umweltprobleme diffusen Ursprungs (NPS) ist die Umsetzung von Agrarumweltprogrammen (AES). In AES erhalten Landwirte Kompensationszahlungen für die Umsetzung von umweltverträglichen Landnutzungsmaßnahmen. AES sind in entwickelten Ländern weit mehr verbreitet als in Entwicklungsländern. Ein zentrales Anliegen bei der Ausgestaltung von Agrarumweltprogrammen ist die Optimierung. Um dies zu erreichen, wurden Modelle entwickelt, die ökonomische und umwelthydrologische Kenntnisse integrieren und dadurch sowohl die ökologischen Auswirkungen als auch die ökonomischen Kosten von Landnutzungsmaßnahmen bewerten. Obwohl solche integrierten Modelle in der Lage sind, Empfehlungen abzugeben, welche Landnutzungsmaßnahmen wo umgesetzt werden sollen, nehmen sie typischerweise eine Planungsperspektive ein und berücksichtigen nicht, dass Landwirte freiwillig entscheiden, ob sie an einem AES teilnehmen. Darüber hinaus ignorieren die Modelle, dass ein Agrarumweltprogramm oft aus mehreren Maßnahmen besteht, aus denen Landwirte eine Auswahl treffen können. Diese Wahlmöglichkeiten müssen bei der Ausgestaltung von AES Berücksichtigung finden.

In diesem Zusammenhang erarbeitet diese Studie einen räumlich expliziten, integrierten Modellierungsansatz, der genannte Unzulänglichkeiten durch die

Kombination von mehreren Komponenten adressiert. Zuerst wurde ein Katalog geeigneter Landnutzungsmaßnahmen für die Untersuchungsregion entwickelt. Zweitens wurden die ökologischen Auswirkungen der Landnutzungsmaßnahmen hinsichtlich der Bodenerosion und Wasserverschmutzung durch Nitrat- und Phosphoreintrag für räumlich heterogene Einheiten durch ein SWAT-Model quantifiziert. Drittens wurden die Landnutzungskosten für die heterogenen räumlichen Einheiten ermittelt. Als Viertes wurde ein Optimierungsverfahren angewandt, das Verhalten von Landwirten berücksichtigt, kosteneffektive AES für die Reduktion von Bodenerosion und Wasserverschmutzung (als N, P) zu entwickeln. Kosteneffektivität bedeutet in diesem Zusammenhang, dass für ein gegebenes AES-Projektbudget die (gewichtete) Sedimentbelastung durch Bodenerosion und N- und P-Belastung minimiert wird.

Die Relevanz dieses Modellierungsverfahrens ist in dieser Arbeit durch seine Anwendung auf das Baishahe Wassereinzugsgebiet, das mit einer Ausdehnung von ungefähr 56 km^2 in der Shanxi Provinz in China liegt, unter Beweis gestellt worden. Die Ergebnisse zeigen, dass der Modellierungsansatz sich als robust erweist, um kosteneffektive AES zu gestalten, worunter die Ermittlung von AES zu verstehen ist, die für ein gegebenes Budget die ökologische Auswirkung durch festgesetzte Zahlungen für Maßnahmen maximieren. Das entwickelte Modellierungsverfahren ist verallgemeinerbar und wirkmächtig hinsichtlich der Anwendung in allen Arten von Wassereinzugsgebieten unterschiedlicher Größe.

Schlüsselwörter: Agrarumweltprogramme · Kosteneffektivität · Integrierte Modellierung · Bodenerosion · Wasserverschmutzung

Contents

Abbreviations

AES	Agri-environment schemes
AnnAGNPS	Annualized agricultural nonpoint source pollution model
BAU	Business as usual
BMPs	Best management practices
CMADS	China meteorological assimilation driving datasets for the SWAT model
DEM	Digital elevation model
Eq.	Equation
ESS	Ecosystem service
EU	European Union
FAO	Food and Agriculture Organization
GFGP	Grain-for-green program
GIS	Geographical Information System
HRUs	Hydrologic response units
HWSD	Harmonized world soil database
N	Nitrogen
NPS	Non-point source
NRCS	Natural resource conservation service
OAT	One-factor-at-a-time
OECD	Organization for Economic Co-operation and Development
P	Phosphorus
PES	Payment for ecosystem services
PS	Producer surplus
PVs	Present values
S	Sediment
SA	Simulated annealing

SHUs	Spatially heterogeneous units
SPAW	Soil-plant-air-water
SWAT	Soil and water assessment tool
TCs	Transaction costs
TMDL	Total maximum daily load
TN	Total nitrogen
TP	Total phosphorus
UK	United Kingdom
US	United States
WTA	Willingness to accept
WTP	Willingness to pay

List of Figures

List of Tables

Introduction

1

1.1 Background and Motivation

Along with the population increase and food security in the past decades, intensified agricultural production has been introduced in China and many other parts of the world. However, intensive agriculture has, in turn, resulted in many severe environmental problems. For example, soil erosion and water pollution caused by cropland cultivation. These problems not only are hard to address but also have broad and profound influences on both upstream farms and downstream waterbodies. The resulted damages include negative effects on long-term sustainability of food production and its resulted economic loss, as well as human health and agro-ecosystem functions (Norse and Ju 2015).

According to the definition of Food and Agriculture Organization, agricultural system could be categorized as sectors of cultivation of crops, animal husbandry, forestry, fisheries, and the development of land and water resources (Ciparisse 2003). Among these sectors, large scale intensive fish farms and livestock production are considered as point sources of water pollution, while the bulk of non-point source (NPS) pollution in agriculture is from cropland cultivation (Brenninkmeyer 1999; Ongley et al. 2010; Norse and Ju 2015). In China, the most concerns of cropland cultivation impact for surface water are sediments (S) due to soil erosion and nutrients of nitrogen (N) and phosphorus (P) (Mateo-Sagasta *et al.*, 2013).

Soil erosion has huge impacts on both on-site and off-site where the sediments generated originally and flowed into in the process and end respectively. On-site impacts result in the decrease of agricultural productivity, because of the soil quality degradation in an irreversible direction (Zhao et al. 2013). The off-site

© The Author(s), under exclusive license to Springer Fachmedien Wiesbaden GmbH, part of Springer Nature 2023
Z. Hao, *An Integrated Modelling Approach to Design Cost-Effective AES for Agricultural Soil Erosion and Water Pollution*,
https://doi.org/10.1007/978-3-658-41340-8_1

impacts are concerned with things happened when soil leaves the field, which pose a much greater threat of environmental issue mainly regarding water quality and muddy flooding (Mullan 2013).

Soil erosion can happen naturally, while scientific evidence has showed that agriculture cultivation leads to greatly accelerated rate of soil erosion (Amundson et al. 2015; Nearing et al. 2017). An experimental study in EI Reno, OK, USA showed that the soil erosion rate of cropland with conventional till, conservation till and no-till are about 173, 68 and 8 times of native grassland respectively (Zhang and Garbrecht 2002). According to Nearing et al. (2017), average soil erosion rate under natural conditions is less than 2 Mg/ha/yr, while under cropland cultivation it is documented as approximately 6.7 Mg/ha/yr in USA and 15 Mg/ha/yr in northeastern China currently. USA has large cropland area, and the conservation practices on cropland have helped the reduction of erosion rate from about 9 Mg/ha/yr to 6.7 Mg/ha/yr between 1982 and 2012 (U.S. Department of Agriculture 2015). China's cropland is mostly small patched, and intensified cultivation in the last century has led to very high erosion rates in critical food production areas, like as high as 23.72 Mg/ha/yr in Jilin province (Nearing et al. 2017).

Cropland soil erosion in China mainly results from cultivation of steep slopes and unsound cultivation practices (Norse and Ju 2015). Since 1999, the government has implemented the Grain-for-Green Program[1] in order to return the sloping cropland to forest or grassland. However, the other cause of unsound cultivation practices are not well emphasized and solved by the Chinese government.

On the other hand, increased anthropogenic nutrient (N and P) pollution is one of the most prevalent concerns for the security of Earth's surface freshwaters (Davis et al. 2019). Eutrophication is the major consequence of nutrient pollution for surface water, and nowadays the main driver of it is the diffuse source of N and P (Le Moal et al. 2019; Beusen et al. 2016). A number of research demonstrate that the mismanagement and overuse of synthetic fertilizer and manure on cropland are the dominant cause of excessive nutrients in waterbodies (Zhou et al. 2016; Norse and Ju 2015; Sun et al. 2012; Norse 2005). N and P are transferred from cropland to waterbodies by surface runoff, eroded sediment, and groundwater discharge (Nyakatawa et al. 2006).

The situation is especially serious in China. The increasing inputs of chemical fertilizers are the principal means in China for attaining high grain production,

---

[1] It is also known as the "Converting Slop Land into Forest", but in this thesis the term "Grain-for-Green Program (GFGP)" is consistently adopted.

consequently China is now the largest consumer of synthetic N and P fertilizers in the world (Li et al. 2019; Sun et al. 2012). However, the N and P use efficiencies in China are relatively low due to sub-optimal management (Zhao et al. 2019). As a result, the excessive fertilizer inputs combined with their inefficient usage have resulted in severe environmental degradation since the 1990s in China (Ju et al. 2009).

NPS water pollution is hard to address. Political regulations are limited and insufficient to achieve desired environmental outcomes, which has induced the increasing proposal of economic instruments acting as effective and cost-effective response to the problem (Sidemo-Holm et al. 2018; Dowd et al. 2008). The common approach is the market-based agri-environment schemes (AES). It refers to voluntary participation programs, where farmers are offered mitigation measures and corresponding payments for them. The aim is to encourage farmers to implement mitigation measures with compensation payments. AES have become widespread nowadays, especially in Europe and the United States, where AES were first promoted and applied and in recent years several billion euros are paid annually through such schemes to farmers for the improvement of environment to their land (Armsworth et al. 2012; Donald and Evans 2006).

Regionally, AES are more applied in developed countries than in developing areas. Under the environment pressure, Chinese government has been developing the concept and practices of AES[2] since the beginning of this century (Pan et al. 2017). The country's efforts for pilot testing of AES programs have been internationally recognized, such as the Grain-for-Green Program[3] with the enormous amount of monetary invested (more than US$69 billion) and the huge area of coverage (32 million households, 15.31 million hectares) (Jin et al. 2017). However, the country' AES programs are also widely criticized for their numerous challenges, like the unreasonable compensation payment without heterogeneity and without scientific design (Yang et al. 2013b; Sun and Zhou 2008; Jin et al. 2017).

Regarding to the topics addressed with AES research, according to a review by Uthes and Matzdorf (2013), the majority of them deal with the ecological effects of AES empirically, while the agri-economic costs of AES are relatively much less emphasized. However, without paying the same attention for the cost analysis, the research cannot provide results that could help for decision making.

---

[2] Relevant programs in China are called "eco-compensation", while in this thesis the term of AES is consistently adopted throughout the thesis.

[3] Many research refer Grain-for-Green Program as AES-like project, as it does not fully follow the principles of AES.

Meanwhile, most of these research are focused on the topic of biodiversity conservation, while for the abiotic resources, such as water quality, relatively less are studied.

A main challenge for AES, which has been attracting increasingly research and practical attention, is the achievement of cost-effectiveness of the program. Cost-effectiveness in AES means to either reach the maximized environmental target under a certain budget or to achieve the minimized budget for the given environmental goal (Balana et al. 2011; Wätzold and Schwerdtner 2005). Two of the points are emphasized for achieving the cost-effectiveness of AES programs. One is the integration with both ecological and economic perspectives involved equally for a holistic picture of AES analysis, which is therefore interdisciplinary research (Ansell et al. 2016; Wätzold and Schwerdtner 2005). Although with the realization in the last decade, the interdisciplinary research remains being a lack, mainly due to the less involvement of economic analysis (Ansell et al. 2016; Uthes and Matzdorf 2013). The other main cause of low cost-effectiveness in AES is poor spatial targeting (Uthes et al. 2010a). Information of spatial heterogeneity of land for both ecological effects and economic costs is important to reduce the producer surplus caused by asymmetric information between ecosystem service (ESS) providers and AES payers (Baylis et al. 2008; Claassen et al. 2008). The more heterogeneous the environmental effects and economic costs are, the greater the cost-effectiveness gains that could be realized through targeted and differentiated payments (OECD 2010; Ferraro 2008).

Regarding NPS water pollution, the integration of ecological and economic analysis regarding agri-environment measures could be combined in hydro-economic modelling[4] (Harou et al. 2009). The components of hydro-economic modelling mainly include an eco-hydrological model for ecological effects simulation, economic costs estimation, and an optimization process[5]. As to the spatial heterogeneity of land, in hydro-economic modelling hydrological units should be adopted as the spatial units, like sub-watershed. These spatial units are shaped according to hydrological principles based on the local topography, which have irregular shapes and heterogeneous sizes.

Previous research has shed some lights on the cost-effective analysis of agri-environmental measures regarding NPS water pollution mitigation with hydro-economic modelling and considering spatial heterogeneity simultaneously. However, the majority of this research were focused from the social planner's perspective, but did not consider the policy implementation of AES. Meanwhile,

---

[4] There are different terms for it in literature, in this thesis this term is consistently used.

[5] The detailed explanation of these components is described in Chapter 2.

most of these research only considered the spatial heterogeneity of ecological effects of mitigation measures, but ignored the same level of heterogeneity for the economic costs of corresponding measures. Regarding the approach for cost-effective analysis, some research attained a ranking of mitigation measures, sets of combination of measures or various scenarios based on their cost-effectiveness ratios (Liu et al. 2019; Cools et al. 2011; e.g. Balana et al. 2015; Lescot et al. 2013). Cost-effectiveness ranking could determine the selection of measures on a certain location, but cannot consider the selection of locations of measures within the whole landscape. A range of research treated the aim of cost-effectiveness as the problem of multi-objective optimization with total economic costs and ecological effects of land use patterns with mitigation measure distribution. These research mostly adopted the optimization method of genetic algorithm or non-dominated sorting genetic algorithm (e.g. Arabi et al. 2006; Maringanti et al. 2011; Rodriguez et al. 2011; Geng et al. 2019; Chen et al. 2015; Dai et al. 2018; Yang and Best 2015). The ecological effects of mitigation measures in these research were heterogeneous and derived from model simulation, while the corresponding economic costs were without heterogeneity. Heterogeneity of both ecological effects and economic costs of mitigation measures was considered in Konrad et al. (2014), with the ecological effects based on experiments and differentiated on soil type and economic costs distinguished for crop categories. Spatial heterogeneity could be considered in more detail to be more realistic, like for economic costs many other factors have impacts on spatial differentiation, including transportation costs, labor costs and others.

All these mentioned research are focused on a planning perspective but not considering at a level of decision-making for AES design, they neglect the role of farmers and the availability of budgets (Uthes et al. 2010a). One exception is Hérivaux et al. (2013), which adopts hydro-economic modelling to identify cost-effective AES combinations for NPS pollution mitigation of groundwater. Although it considers the voluntary participation of farmers in AES programs, it involves only four AES (each has only one measure) and there is no spatial heterogeneity considered for the pursuit of cost-effectiveness.

To fill these gaps, this thesis develops a methodology which could design cost-effective AES with an integrated hydro-economic modelling procedure, considering spatial heterogeneity of both ecological effects and economic costs of mitigation measures in the same detail level. The cost-effectiveness here refers to the achievement of maximum ecological effects of AES under certain limitations of AES budgets. The study focuses on how to design compensation payments to farmers in AES to achieve the aim of cost-effectiveness. It could show the optimal distribution of various mitigation measures in a landscape for NPS water pollution in AES programs with given budget, which is resulted from a voluntary participation of farmers. The study could provide results at the level relevant to decision-making, with both the role of farmers and the availability of AES budgets being considered. Meanwhile, there are 12 mitigaiton measures and 50 spatially heterogeneous units. This demands a complex numerical optimization method with computer operation.

The novelty of the study is that it is an explicit research targeting at cost-effective AES design with the consideration of measure implementation of famers instead of from a social planner's perspective, as well as integrating interdisciplinary components of ecological effects and economic costs in a hydro-economic modelling procedure and considering the heterogeneity of them in the equally spatial level. With this, the thesis focuses on developing a method for cost-effective AES design on environmental problems of NPS water pollution from agricultural cropland. The method is demonstrated on a typical agricultural landscape with the major land use being cropland and NPS water pollution from cropland being emphasized in the watershed. To realize this novelty, the study is highly complicated with multiple disciplines involved and corresponding scientific techniques being applied in the process. A similar approach is developed by Wätzold et al. (2016). However, it targets the counterpart environmental problem of biodiversity conservation, with which the explicit methods are totally different and their spatial units for the consideration of heterogeneity are uniform squares. In contrast, except the different methods applied in this study, with hydro-economic modelling the spatial units of heterogeneity are irregularly shaped and various in sized. This leads to new challenges as well as new findings, especially in the process of optimization during the study.

1.2   Objectives and Approach

The general aim of this thesis is to develop a method for the design of cost-effective AES regarding NPS water pollution mitigation of soil erosion and water pollution from cropland to waterbodies. The specified mitigation targets of pollutants are sediments due to soil erosion on cropland and N and P caused mainly by excess fertilizer application during cropland cultivation. The method is developed through the cooperation and coordination of interdisciplinary components, which forms an integrated hydro-economic modelling procedure. The involved knowledge and techniques regarding cross-discipline include agronomic process (mitigation measures), hydrological modelling (mitigation effects), economic analysis (opportunity costs), and heuristic optimization method (cost-effectiveness).

To develop this method, some basic settings and assumptions are made. First, a study region in real is selected, where this study assumes some AES programs are planned to be implemented. These AES differ with each other over their set budget levels, specific mitigation targets, and the inventories of available mitigation measures. For a certain AES with exact budget amount, mitigation target and inventory of mitigation measures, the study can get a result of its corresponding maximum total mitigation effect in the study region, based on the determination of which measures should be offered to farmers and how much should be their corresponding compensation payment. The method developed here is basically to show how to make this kind of determination. Temporally, these AES programs are assumed to be implemented for a period of five years[6], with the beginning year in 2018 and ending year in 2022. The amount of compensation payment to farmers for each offered measure should be annually equal. Both the mitigation effects and economic costs of measures are assessed in terms of average annual results.

Second, spatial heterogeneity is emphasized in this study for the pursuit of cost-effectiveness. The study region is divided into many spatially heterogeneous units (SHUs). Among these SHUs, the ones occupied with cropland are the focus, each of which is treated as an individual farm and is assumed to be charged by a representative farmer. In an integrated hydro-economic modelling procedure, the SHUs for mitigation effects simulation and for economic costs calculation should be consistent. In this study, eco-hydrological model of SWAT (Soil and Water Assessment Tool) is adopted to divide the study region into many heterogeneous HRUs (hydrologic response units), including 50 HRUs occupied with cropland. These 50 HRUs are morphologically the 50 SHUs of cropland for mitigation effects simulation and opportunity costs estimation, corresponding to the assumed farms. For each of these 50 cropland SHUs, in an AES there is at most one mitigation measure being selected by the corresponding farmer in this study. This is based on the consideration that farmers might not be willing to apply more than one measure at a time, also it could avoid the problem of mutually exclusive measures being applied on the same land plot, as described in Konrad et al. (2014).

---

[6] Contracts of AES programs typically run for a five-year period, as mentioned by Hérivaux et al. 2013 and Uthes et al. 2010a.

Third, farmers are assumed to be benefit-maximizers, who pursue the maximum benefits under available options. In AES programs, farmers voluntarily decide on whether to participate as well as selecting which measures and where. Farmers' selection behavior is assumed to be based on the comparison of net economic benefits resulted from the payments and costs of measures. Several measures might be offered to farmers with uniform payment for each one. Combined with the information of costs of each measure in each SHU, farmers could make their option.

These settings and assumptions are the basis for the research aim of method development. The developed method is demonstrated through its application in a real study region, which is called the Baishahe watershed located in Shanxi province in China. Based on the situation that China invested large on AES programs but with less cost-effectiveness, taking the study region in China would be meaningful and demanded. the Baishahe watershed is a typical agricultural area focused on cropland cultivation, where other agricultural activities are less happened and there is no business firms for big contribution of point source pollution. While, the developed method in this study is generic, it could be broadly applied on other agricultural areas with cropland production.

Substantial amount of and very detailed data are needed in this study. Both primary data and secondary data are collected for different parts of the study, with various data processing methods being applied. But still, data limitation happens in this study due to time reason and some data's unavailability. It makes the application of the results of the study in practice in the study region to be unconvincing, but it is reasonable to give some general suggestions for the local. However, the emphasis of the study is method development, with which the effort for the data in the thesis is adequate.

To achieve the research aim, several specified objectives could be clarified:

(I) To identify appropriate mitigation measures to form the inventory of measures for AES design in this study. These measures are aimed to reduce the load of either sediments or total nitrogen (TN)[7] and total phosphorus (TP)[8] from cropland to waterbodies in the watershed. The identification of

---

[7] The sum of all kinds of forms of existed nitrogen in water, including organic N and mineral N (like NO_3, NH_4, NO_2).

[8] The sum of all kinds of forms of existed phosphorus in water, including organic P and mineral P.

measures should be based on the information of the study region's characteristics, as well as other principles, such as coordinating with the followed tasks of mitigation effects simulation and opportunity cost calculation.

(II) To obtain the quantified mitigation effects of each identified measure in each SHU in the study region. In this study, simulation results with an appropriate eco-hydrological model are preferred. SWAT model is selected in this study. With SWAT, the mitigation effects of each measure in each HRU are wanted at the total outlet of the watershed instead of in the HRU level as in many other studies (e.g. Arabi et al. 2006; Maringanti et al. 2011). The purpose of this is to consider also the influence of in-stream process for pollutant transport. This requires that when doing the simulation of one measure in one SHU, other SHUs should keep business as usual (BAU) situation.

(III) To obtain the quantified economic costs incurred to farmers when they implemented each identified measure in each SHU. If the compensation payment of a measure is higher than the cost of it, farmers could earn a producer surplus (Wätzold and Schwerdtner 2005). Cost-effective AES design pursues to reduce the amount of this kind of producer surplus as more as possible. For doing this, it is crucial for AES designers to get the information of the costs of measure implementation, which is clear to farmers. In this study, formulas of cost calculation for each identified measure need to be formed based on the analysis of involved cost activities and categories during every step of measure implementation. The spatially heterogeneous factors regarding cost calculation need to be considered, which make one specific measure has different costs in different SHUs. Data for cost calculations need to be collected mainly from local farmers.

(IV) To figure out a way to make the interdisciplinary components of work mentioned above to be integratable in a manner that they are coordinated and cooperated with each other. As described by Brouwer and Hofkes (2008), there are many operational challenges to link hydrological and economic systems, including their different conventions for the dimension of spatial units and time scales. This kind of technical units need to be consistent during the hydro-economic modelling procedure. Except this, measures are kind of bridge between hydrological model and economic analysis. Therefore, for the identification of measures it needs to consider both the simulation capability of hydrological model and the feasibility for the calculation of economic costs of the measures. In this process, reversed influences between those components and the reduction of these influences also need profound consideration.

(V) To do the simulation of farmers' selection behavior in AES programs, and at the same time to do the optimization for getting the cost-effective AES. Simulation of farmers' selection behavior means that to mimic farmers of which measures they would select for which SHU with the offered measures and corresponding payments, based on that they are benefit-maximizers. In this process, the payment for each measure is given, based on which and the costs of measures farmers will select for a SHU at most one measure that has the maximum positive net economic benefit (payments minus costs of measures). Optimization is to identify an AES consisting of a set of measures and related payments, which induces farmers to select measures in a way that the resulting land use pattern of an AES generates the maximum total mitigation effects for a given budget level for AES. It involves an objective function with the quantified total mitigation effects being the objective and a certain amount of budget as the limitation. The simulation of farmers and optimization process need to be combined, with the complexity of which a specially programmed optimization modelling procedure is needed to finish this task. A computer scientist is demanded to assist and cooperate to finish the programming of this modelling procedure. As a heuristic optimization method, simulated annealing (SA) is adopted as part of the tools of this modelling procedure.

1.3   Structure

The structure of this thesis is arranged as the following. In Chapter 2, some background knowledge and the status quo of related research are demonstrated. These include the theoretical and practical development of AES, the overall situation of AES in China, and the development and the status quo of the interdisciplinary research of hydro-economic modelling procedures, including the commonly adopted tools for its components.

Chapter 3 gives some information about the study region for this thesis, covering mainly three aspects. Firstly, the general characteristics of the area is described, including physical environment, location and climate, social structure, and situation of agricultural production. Secondly, the pollution situation and the main sources of pollution in the study watershed are analyzed. Thirdly, the situation on the policy and practices for soil and water conservation in the area is introduced.

Chapter 4 presents the overall framework of the hydro-economic modelling procedure in this study. It is an introduction for the general methodology of this study, with a graphically technical flowchart to show the position of different interdisciplinary components, their corresponding functions, their connections between each other, and how can they be integrated to achieve the research aim. Meanwhile, a way to make these interdisciplinary components to be consistent and coordinated in a systematic study, as required in one of the specified objectives in above, is demonstrated.

Based on the overall framework in Chapter 4, each of the individual components involved are focused in each following chapter respectively from Chapter 5 to Chapter 8. Chapter 5 aims at identifying the appropriate mitigation measures for load of pollutants, i.e. sediment, N and P, from cropland to waterbodies. The identification of measures should be based on some principles, which are then described and explained with the reason. The final identified mitigation measures for this study are described with their corresponding functions, features and operational procedures.

Chapter 6 focuses on obtaining the mitigation effects of each identified measure in each SHU. It starts from identifying the appropriate eco-hydrological model for this study, which is SWAT. Then the SWAT model introduction and overview are stated, based on which the setup of SWAT model for the study region is processed, including data preparation, model configuration, calibration and validation. Following this is the model simulation of each identified measure for getting their corresponding mitigation effects in each SHU in the study region. Afterwards, the simulation results for different measures in different SHU are analyzed.

Chapter 7 focuses on the analysis and calculation of costs happened to farmers if they implement each of the identified measures in each of SHUs in the study region. To do this, first the general category of conservation costs and afterwards the components of compensation payments are briefly analyzed. Based on this, one significant component, i.e. the production costs of measures, is focused for the cost calculation of all identified measures, along with the corresponding formula development, data collection and processing, and results analysis.

Chapter 8 describes the optimization modelling procedure combining the process of simulation and optimization of AES. The assumptions, principles and logic of the simulation of AES, i.e. how to mimic farmers' selection behaviors in an AES program are explained. For optimization of AES, the relevant objective function is demonstrated, together with the introduction of the adopted heuristic optimization method of SA. The optimization modelling procedure results in a software tool, which is programmed by a computer scientist specially for this study.

Chapter 9 shows the results of various cost-effective AES in this study, each of which is corresponding to a special land use pattern with a set of payments of measures, a previously set budget level, total mitigation effect and others. Cost-effective AES are obtained through the combination of the intermediate results from Chapter 6 and 7, i.e. the quantified mitigation effects and costs of each identified measures in each SHU, and the input of the combined data into the programmed modelling procedure. It is flexible to get various cost-effective AES with different requirements for mitigation targets, budget levels, and available measures.

Chapter 10, as the end, makes the conclusion and discussions for this study, including the major novelty of the study, shortcomings in some perspectives, policy recommendations of the study region, possible improvement for the study, and outlooks for future research.

Background Information

2

This chapter gives some background knowledge on the research and practical status of AES, including its principles and mechanism, as well as origin and development. Along with this, the overall situation of AES development in China is described, together with the correspondingly special features and challenges of AES in the country. Meanwhile, the status quo of hydro-economic modelling procedure is briefly reviewed.

2.1 Overview of AES

2.1.1 Principles and Mechanism

AES, according to the explanation of OECD[1], are payments to farmers and other landholders for addressing environmental problems and the provision of environmental service (Van Tongeren 2008; Tarek 2010). To understand the mechanism of AES in a further step, it is necessary to introduce the concept of payment for ecosystem services (PES). There are some distinctions between AES and PES, although they might be subtle. Generally, AES target at specific farming practices on agricultural land and are adopted mainly in developed countries, while PES are aimed for broader ESS problems and implemented on a large scale in developing countries (Daniela 2011; Le Coënt 2016). Although these differences, AES and PES have the same logic and principles behind them, even AES are discussed

---

[1] Organization for Economic Co-operation and Development.

© The Author(s), under exclusive license to Springer Fachmedien Wiesbaden 13
GmbH, part of Springer Nature 2023
Z. Hao, *An Integrated Modelling Approach to Design Cost-Effective*
AES for Agricultural Soil Erosion and Water Pollution,
https://doi.org/10.1007/978-3-658-41340-8_2

by some as a form of PES or PES-like programs (Wynne-Jones 2013). Different versions of the definition of PES are given in the literature, here a commonly accepted one described by Wunder (2005, p. 3) is considered in order to explain the mechanism of PES and AES. It is:

> "(I) a voluntary transaction where
> (II) a well-defined ESS (or a land-use likely to secure that service)
> (III) is being 'bought' by a (minimum one) ESS buyer
> (IV) from a (minimum one) ESS provider
> (V) if and only if the ESS provider secures ESS provision (conditionality)."

From this definition, it is implied that PES programs are in effect on attempt to put into practice the Coase theorem (Engel et al. 2008). "A well-defined ESS" in the definition of PES is actually treated as a kind of commodity for an economic transaction. Compared to PES, in AES programs this kind of commodity for transaction is usually some well-defined farming practices[2] which are targeted to reduce environmental problems, such as identified mitigation measures in this study. To make this kind of special commodity and transaction to be feasible, it requires mainly five aspects in an AES program to be clear and satisfied. As described in the definition, these aspects refer to the voluntary participation of stakeholders, specific identified farming practices as the "commodity", buyers and sellers of this "commodity", and transaction between them with the condition of farming practices being implemented and compensation payment being provided.

Based on this, the basic principles of AES mechanisms could be illustrated as in Figure 2.1. It takes the example of water pollution problem resulted from upstream conventional farming activities. Under the conventional farming situation, farmers could get their maximum economic benefits from agricultural land (the amount of this benefit is illustrated as yellow box in Figure 2.1). However, under this situation, there might be pollutants from agricultural land, which bring damages to the downstream populations (the amount of cost caused from this damage to the downstream people is illustrated as blue box in Figure 2.1). A way to reduce the load of pollutants from upstream to downstream is to make farmers to implement some mitigation measures in their farming process. While,

---

[2] Many AES studies or programs are directed on result-oriented, which means to take the final provided ESS on the agricultural land resulted from farming practice implementation as the condition of compensation payment. This study takes only the implementation of farming practices as the payment condition, and the AES mechanism is explained based on this situation.

as benefit-maximizers, farmers have no willingness to do this without any compensation payment, as usually the uptake of mitigation measures would lead to the reduction of their economic benefits from agricultural land (the amount of economic benefit when implementing measures is illustrated as purple box in Figure 2.1). To solve this problem, it is reasonable to get some compensation payments from the downstream populations (as the ESS beneficiaries) to farmers (as the ESS providers) in order to drive them to implement the mitigation measures. The amount of compensation payment should be more than the loss of the farmers (a part of yellow box) if they implement the measures and less than the total cost of downstream populations (blue box). In this way, in an AES program upstream farmers can make sure that their final economic benefits (illustrated as the sum of green and purple box in Figure 2.1) are not less than before, while downstream people could reduce their damage from pollution.

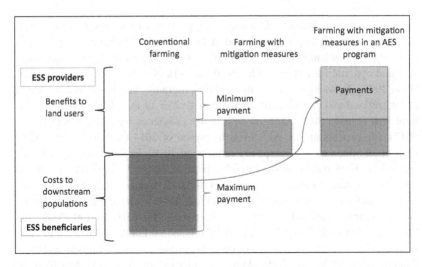

Figure 2.1 The principles and mechanism of AES Source: Modified from Engel et al. (2008) and Pagiola and Platais (2007)

2.1.2 Origin and Development

Generally there is more concern for farmland conservation in Europe relative to other parts of the world (Batáry et al. 2015; Wilson et al. 2009). The European

Union (EU) agricultural policy first explicitly addressed the environmental problem from agriculture in a Green Paper published in 1985 (Kleijn and Sutherland 2003; Commission of the European Communities 1985). In the same year, the reform of the EU agricultural policy (EU Regulation 797/85) included a novel set of measures for environmental protection (Kleijn and Sutherland 2003), which was conceived as a mechanism to compensate farmers for less intensive management of Environmentally Sensitive Areas. This could be the origin of most European AES (Batáry et al. 2015). According to Dobbs and Pretty (2008), the program launched in the United Kingdom (UK) in 1986, the Environmentally Sensitive Areas scheme, was the first AES program in the EU. In 1992, EU Regulation 2078/92 introduced agri-environmental programs as a supplement to the Common Agricultural Policy instruments (Schomers and Matzdorf 2013), which made AES became compulsory for all EU members according to environmental needs and potential (Batáry et al. 2015; Kleijn and Sutherland 2003; Dobbs and Pretty 2008).

Along with the origin of AES, many AES programs are initiated. In EU, Common Agricultural Policy was reformed in 1999 and 2003 to better improve AES application, maintaining the nature of AES being compulsory for member countries and optional for farmers under the framework of rural development program (Pion 2007). Nowadays, a diversity of AES programs exist in EU countries and in Switzerland and Norway, and approximately 25% of the EU's utilized agricultural area is under AES contracts with farmers (Science for Environment Policy 2017). EU expenditure on AES over the period of 2014 to 2020 is expected to be total 25 billion euros (European Commission 2019). Except Europe, AES programs have been widely adopted in other parts of the world, especially in OECD countries outside of Europe (Vergamini et al. 2015; Uetake 2013). The examples include the Conservation Reserve Program and the Environmental Quality Incentives Program in the United States (Lambert et al. 2007; Baylis et al. 2008), the National Soil Conservation Program and the National Landcare Program in Australia (Hajkowicz 2009), and many other programs in Japan, Canada and New Zealand (Uetake 2013). While, AES programs in the worldwide display some general trends of differences according to the cultural and geographical context. For example, some regions prefer the AES with less intensive farming practices (like in EU and USA), others prefer payments for land retirement (e.g. EU, USA and Japan), and the preference of structural changes of land use in general (Pion 2007).

AES programs are carried out through the implementation of agri-environmental measures. Each of these measures has at least one environmental objectives, which include biodiversity conservation, landscape protection, soil, water and air quality enhancement, climate change mitigation (United Nations Convention to Combat Desertification 2017). According to a review on AES studies by Uthes and Matzdorf (2013), the majority of AES research concentrates on biodiversity conservation as well as the ecological effects of that, while abiotic resources are relatively less targeted. However, more recently there is increasing research with the emphasis of AES on the goals of improving and maintaining all different kinds of ESS, such as water quality, pollination and biocontrol (Batáry et al. 2015; Ekroos et al. 2014). Meanwhile, increasing attention has been attracted to the research which design cost-effective AES with quantitative and spatially heterogeneous information (e.g. Bamière et al. 2011; Uthes et al. 2010b; Drechsler et al. 2007a; Wätzold et al. 2016).

2.2   Overall Status of AES in China

2.2.1   Initiation and Development

Several disasters happened in China with large areas and severe consequences in the end of last century, which induced the government start to consider sustainable development and triggered the implement of all kinds of environmental protection projects (Pan et al. 2017; Sun and Zhou 2008). In summer 1998, massive floods happened in major river basins of China, which resulted that several thousands of people lost lives, 14 million people to be homeless, and US $20 billion of economic losses. The severity of the flood is mainly due to the extensive logging and sloping land cultivation in the past years. In 2000 spring, severe sandstorms happened in northern area of China with unprecedented scope and intensity, and repeated in 2002 spring. The main reasons for both of the events are the increasing serious soil erosion and desertification in the west and north areas of China (Sun and Zhou 2008).

Because of these, National Forest Law of China was revised in 1998 for establishing a national fund to compensate ecological benefits of forests, which is the institutional origin of PES in China (Pan et al. 2017). Since then, a range of environmental protection projects are operated in China with large spatial areas and long temporal scopes funded mainly by central government. These include the Grain-for-Green program, the Natural Forest Protection program, the Returning Grazing Land to Grassland project, Beijing and Tianjin Sandstorm Source Control project, Ecological and Environmental Protection program in Three-river Source, Karst Rocky Desertification Control project in southwest China, and the payments for ecological benefits of non-commercial forest.

However, many of these programs are not PES and AES but only conservation projects as the investment are mainly for conservation engineering but there is no payments and subsidies to relevant farmers. Such as the Beijing and Tianjin Sandstorm Source Control project, which have spent about 5 billion euros in the first stage (2001–2012) and have invested 11 billion euros for the second stage (2013–2022) for basic conservation engineering (ChinaDaily 2012), e.g. planting trees and seeding in grassland, but without any subsidies to farmers, mainly acting for the aim of transiting to sustainable development model. The Grain-for-Green program is typically considered as PES, while debate exists as the implementation of the program in reality is not totally consistent with the principles of PES definition. The investment of Grain-for-Green program not only includes the compensation payments to farmers but also covers the basic conservation engineering, and the participation of farmers in some parts of areas are not completely voluntary. This kind of programs, which involve compensation payments to farmers but not exactly conform to the principles of PES, are termed in some research as PES-like programs (Yang et al. 2013b). Other PES-like programs include the Natural Forest Protection program and the Returning Grazing Land to Grassland project. According to Pan et al. (2017), the first real PES scheme is the one established in 2001, i.e. payments for ecological benefits of non-commercial forest (Pan et al. 2017).

Along with the development of these practical programs, a range of government policies are gradually developed at national, provincial and ministerial levels to guide the establishment and improvement of PES schemes (Pan et al. 2017; Wu et al. 2019). Based on this, regarding cultivated land protection, the first official document in China aimed to do the protection-compensation mechanism by promoting AES is released in 2008, with the continuous provisions being enacted in the following years (Zhu et al. 2018). Guided by these policies, many provincial governments have shown increasing interest and some have conducted their experiments with AES programs in their major cities, such as in Chengdu, Suzhou, Dongguan, Guangzhou, and Shanghai (Cai and Yu 2018; Zhu et al. 2018). Meanwhile, more and more attention have be attracted to scientific researchers regarding different perspectives of PES and AES development in China (Zhu et al. 2018).

2.2.2 Relevant Programs

By now, China has applied many programs which compensate farmers or related households for the aim of environmental protection. Although efforts are made to learn from overseas to do the experiments of AES applications, it is hard to find programs in China which are totally consistent with all the principles of AES. Here the identification of whether a program is AES will not be made, but four relevant programs are described in detail, aiming to introduce the situation of the environmental protection-compensation mechanism programs in China. Two of these programs are fully established and implemented in the national scope, i.e. the Grain-to-Green program and grassland conservation program; and the other two programs are only implemented in the watershed, i.e. farmland protection fund in Chengdu and farmland eco-compensation in Suzhou. These programs are shown in Table 2.1, with the corresponding description for each in below.

Table 2.1 Relevant programs in China and corresponding features

Scope	Programs	Starting year	Payees	Conditions required
national	Grain-to-Green program	1999	farmers	convert steeply sloping cropland ($\geq$ 25 degrees) to forest or grassland
	Grassland Ecological Protection Subsidy and Reward program	2011	herders	grazing abandoned, sustainable grazing practiced
local	Farmland protection fund program (Chengdu)	2008	farmers	cultivated lands cannot be abandoned or destroyed
	Farmland eco-compensation program (Suzhou)	2010	farmers	continue to grow rice in rice-growing areas

Source: Content adapted from He and Sikor (2015), Shao et al. (2017), Zhu et al. (2018), and Cai and Yu (2018)

The Grain-to-Green program was initiated in 1999 as a pilot program and was launched in the national scope since 2002, with the major aim of combating deforestation, ecological degradation, over-cultivation of sloping land and soil-erosion (Liu and Wu 2010). The mainly promoted activity of the program is to convert the steeply sloping cropland and marginal land, which are the major contributors of soil erosion in the country, to forests or grassland (Bennett 2008; Zhi et al. 2002). It is considered as one of the largest PES schemes in the world, in terms of area coverage, number of land managers, and amount of financial supports (He and Sikor 2015). The payment is fully from central government in terms of both cash and in-kind. Farmers who convert degraded or highly sloping cropland back to either "ecological forests"[3], "economic forests"[4] or grassland are compensated with either annual in-kind subsidy of grain, or cash subsidy, or free seedling. However, with the big area of the whole nation the compensation standards are mainly differed in two classes in space, like the grain subsidy is 2250 kg/ha in the Yangtze River Basin and 1500 kg/ha in the Yellow River Basin. The design, implementation and means of payments of the program are all in the form of top-down approach, which have induced lots of shortcomings. These include the lack of investigation and consideration of various local conditions, neglect of local innovation and variation, unreasonable payment levels, involuntary participation of many farmers (He and Sikor 2015; Bennett 2008).

Grassland Ecological Protection Subsidy and Reward program was started in 2011 along with the corresponding policy, aimed to relieve overgrazing and promote sustainable stock farming in arid or semi-arid northern area of China. The new round of policy for this program was released in 2016 for the nest five years, which expanded the coverage area of the program from eight provinces to 13 provinces located in northern and northwestern China (The office of Ministry of Finance and the office of Ministry of Agriculture 2016). The program is solely funded by the central government in China. Compensation payments are targeted to herders with the conditionality of either abandoning grazing in severely degraded pasture area, or reducing the number of grazing sheep and doing the rotational grazing in sustainable grazing area. The compensation quota from central government to provincial governments is annually per hectare US $ 17.6 (RMB 112.5) for grazing abandoned area and US $ 5.8 (RMB 37.5) for sustainable grazing area (Pan et al. 2017; The office of Ministry of Finance and the office of Ministry of Agriculture 2016). However, the specific payment criteria for herders in different provinces are different with different specific conditionality requirements. Research showed that herders' willingness to accept is more than the offered payments in some areas (Zhen et al. 2014).

In order to curb farmland loss, retain rural landscape, and to protect farmland resources, Chinese government attempted to adopt monetary or insurance compensation programs to support farmland conservation (Cai and Yu 2018). At the beginning of 2008, central government in China released the No. 1 Central Document[5], with the announcement of "draw boundaries for permanent prime farmland and establish protection-compensation mechanisms" (Zhu et al. 2018, p. 506). The policy aimed to effectively protect both quantity and quality of

---

[3] Defined by the State Forestry Administration as timber-producing forests.

[4] Orchards or plantations of trees with medicinal value.

[5] A government document similar to the Farm Bill in America.

cultivated land under the social situation of rapid progress of urbanization in China (Xiao et al. 2019). After the first release in 2008, the policy is continuously updated each year between 2009 and 2016 except 2011. Under these policy supports, many major cities started to conduct their experimental programs for farmland conservation with compensation payment to farmers. Among these cities and corresponding programs, two typical compensation modes are noticed, i.e. insurance compensation like farmland protection fund in Chengdu and monetary compensation like farmland eco-compensation in Suzhou (Table 2.1).

Farmland protection fund program in Chengdu was started in 2008, which was the first incentive-based farmland project in China for the protection of cultivated land area from decreasing. The city government set up a fund as the source of compensation payments, with 90% of payments in the form of farmers' pension insurance, 10% of payments being used for land transfer guarantee funds and agricultural insurance (Zhu et al. 2018). The conditionality of the payment is that framers need to ensure that the arable land cannot be abandoned without cultivation or destroyed for other usages, otherwise farmers not only loss the payments but also get economic penalties (Xiao et al. 2019). Compensation payment is based on farmland quality, which is 6000 RMB (779 euro)/ha/year for prime farmland[6], and 4500 RMB (584 euro)/ha/year for general farmland[7].

Farmland eco-compensation program was launched in Suzhou in 2010, with the aim of encouraging farmers in the rice-growing area with the economic compensation for them to continue to plant rice (Xiao et al. 2019). The local government raised the fund from various sources, including local finance, special subsidies from upper level government, and social donation (Zhu et al. 2018). Participants of farmers get the compensation in the form of entire cash payment. Compensation payment is based on farmland quality, location and scale, which for prime farmland is 3000 RMB (390 euro)/ha/year where the continuous rice-growing area is between 66.667 and 666.667 hectare, and 6000 RMB (390 euro)/ha/year where the continuous rice-growing area is above 666.667 hectare.

2.2.3 Characteristics and Challenges

Although with large area of implementation and numerous amount of investment, the economic compensation for environmental protection programs in China are

---

[6] Farmland with high quality.

[7] Agricultural production land except for prime farmland.

still in the stage of learning form mature countries, and attempting with experiments and explorations (Xiao et al. 2019). Many implemented programs are based on the mechanism of environmental protection by economic compensation, while most of them cannot be identified as real AES due to their variation on the principles of AES. A market-based AES program in China with farmers' totally voluntary participation is hard to established.

The massive participation of government with top-down approach has induced all kinds of problems, and is considered as the major reason for poor performance of AES programs in China (Shang et al. 2018). Top-down approaches are typical in China for AES related programs. Basically all relevant programs in China are applied under some policy regulations, which give guides on all perspectives of the programs, such as program location, modes, payment standards. However, top-down approach involves multiple levels of governments, resulting in all kinds of severe problems, like corruption, incomplete information and delay of information and compensations (Shang et al. 2018; Bennett 2008). Regarding the design and implementation of AES related programs, higher level of government often make decisions for detailed practices, while these decisions often lack sufficient investigation for the local situation (Benett and Carroll 2014; Sun and Zhou 2008). Participation of farmers is often mandatory when government has designated land for an AES related program (Zhen et al. 2014). One reason for this is that in China farmers only have land use rights but not property. Due to the dissatisfaction of farmers, the effectiveness of achieving environmental goals and the cost-effectiveness of AES programs in China is reasonably doubtable.

Along with the large coverage area of most programs in China, the compensation payment levels for each program are quite generous and mostly unreasonable for suiting to the local conditions. For example, regarding the Grain-to-Green program, there are only two kinds of payment levels for the whole country scope, which resulted in the unfair effects of either "excess compensation" in some areas or "less compensation" in other areas. Another case is the PES for grassland conservation program. According to Zhen et al. (2014), the payment levels are less than the amount of average willingness to accept of farmers in the watershed, resulting in negative effects for the effectiveness of program implementation. This shortcomings of PES in China is often criticized by researchers (Sun and Zhou 2008; He and Sikor 2015; Bennett 2008). It is mainly due to that there is no formal pre-program analysis for the participant opportunity cost being conducted. Moreover, the payment levels are generally constituted by government officials in China instead of experts (Sun and Zhou 2008).

2.3   Overview of Hydro-Economic Modelling

2.3.1   Hydro-Economic Modelling

To pursue the cost-effectiveness of AES, integration of quantified ecological effects and economic costs in an optimization framework is needed. Specified for water resource problems, the combination of eco-hydrologic models and economic cost analysis is required to form an integrated procedure for cost-effectiveness analysis, which could be termed as hydro-economic modelling (Harou et al. 2009). According to Essenfelder et al. (2018), eco-hydrologic models here refer to the kind of hydrologic models that are capable of simulating the interactions between water and ecosystems, like nutrient cycling, vegetation and crop growth, and provision of ecosystem services. The difficulty regarding hydro-economic modelling is the cooperation and coordination of interdisciplinary components based on their interrelationships in a study, including hydrology and ecology (ecological effects), economics (costs), and engineering and agronomy (technical details of measures) (Cools et al. 2011; Cai et al. 2003). Challenges also include the consistence of spatial and temporal scale from different disciplines (Brouwer and Hofkes 2008). In hydrological models, the geographical unit is usually the watersheds and basins, while for economic analysis it is normally administrative boundaries, such as province, county. For time scale, hydrological model often refers to days, months, or seasons, while cost analysis are usually longer, like years.

There are generally two main approaches to integrate individual components or models into hydro-economic modelling, i.e. modular approach and holistic approach (Harou et al. 2009; Brouwer and Hofkes 2008). In modular approach, a connection is built between independent sub-models; while in holistic approach all inputs and outputs are endogenously calculated within a single systematic model. Regarding applying integrated eco-hydrological modelling for the aim of cost-effective combination of AES measures to reach water quality improvement, the general steps are summarized based on studies in Europe related to the European Water Framework Directive (Balana et al. 2015; Martin-Ortega et al. 2015). The process is identifying water quality targets and mitigation measures, baseline application of hydrological model to reproduce recent hydrological conditions and responses, effectiveness simulations of AES measures, cost estimation of AES measures, and optimization and cost-effectiveness assessment in sequence.

Regarding individual components in the hydro-economic modelling, previous studies on cost-effectiveness analysis with NPS water pollution have shed some lights. The abatement measures for agricultural NPS water pollution mainly refer to best management practices (BMPs), which were first proposed by the U.S. Clean Water Act and have been widely applied and proven to be effective for agricultural NPS control (Chen et al. 2015; Dai et al. 2018; Geng et al. 2019). Effectiveness simulations for these BMPs could be performed through the eco-hydrologic models, among which the Soil and Water Assessment Tool (SWAT) acts as the mostly adopted one in the literature mentioned above, others include Annualized Agricultural Nonpoint Source Pollution model (AnnAGNPS) (Qi and Altinakar 2011) and INtegrated CAtchment model of phosphorus dynamics (INCA-P) (Martin-Ortega et al. 2015). Regarding cost estimation for BMPs, the considered components have establishment or construction costs, maintenance costs, and foregone profits as opportunity costs (Arabi et al. 2006; Ahlvik et al. 2014).

Within hydro-economic modelling, spatial heterogeneity is emphasized, as is has been identified as a primary concern for cost-effectiveness of AES (Uthes et al. 2010a). Most of literature with hydro-economic modelling for NPS water pollution focused on the spatially heterogeneity of the ecological effects of abatement measures in a watershed (e.g. Geng et al. 2019; Maringanti et al. 2011; Arabi et al. 2006). On contrast, although the spatial variability of the economic cost analysis also plays significant roles for achieving the aim of cost-effectiveness, it has attracted relatively much less attention in the literature.

2.3.2 Simulation and Optimization

With the simulation and optimization methods, the integration of individual independent components in hydro-economic modelling, i.e. eco-hydrological modelling and economic analysis is carried out. Simulation and optimization refer to different kinds of problems, i.e. "what if" and "what is best" respectively, and they can be applied together for a study or separately (Harou et al. 2009). Simulation means to simulate decisions or behaviors of water resource operations of relevant stakeholders based on predefined rules; while optimization is to attain the optimized water allocation through a mathematically stated objective function subject to some constraints (Pulido-Velazquez et al. 2008).

Optimization in the modelling procedure refers to apply some optimization method to obtain the best sets of payments for measures which could lead to maximum mitigation effects duo to the farmers' selection decisions. The objective for the optimization is total mitigation effects, with the constrains of budget limitation for all mitigation measures. Regarding the optimization method for prioritizing selection and placement of mitigation measures for more efficient water pollution mitigation planning, heuristic search algorithm have be widely applied and showed well performance. Different optimization methods with heuristic search algorithm include generic algorithms, tabu search, and simulated annealing (Geng et al. 2019). Genetic algorithms is applied in the majority of research for watershed abatement measures allocation, such as Dai et al. (2018), Arabi et al. (2006), Maringanti et al. (2011), Rodriguez et al. (2011). Compared to generic algorithms, the other two methods are quite less applied in the same field, such as with method of tabu search in Qi and Altinakar (2011) and method of simulated annealing in Efta and Chung (2014).

Method of Genetic algorithms is based on the principle of evolutionary in biology, with the natural optimal selection of population chromosomes resulted from reproduction, recombination, crossover and mutation of offspring. Tabu search is a meta-heuristic method to avoid the local optimization with its flexible memory design, which create a balance between search intensification and diversification. Simulated annealing is also meta-heuristic, which adopts the principle of annealing in metallurgy in order to jump out the trap of local optimization and then to approximate the global optimum.

For getting the cost-effective AES design with a hydro-economic modelling procedure, although it is very complicated, the algorithm of simulation process and the application of optimization method need to be integrated into one process as the optimization modelling procedure. The optimization modelling procedure should be computerized duo to its complexity of calculation. Only with this optimization modelling procedure, the cost-effective AES design with numerous SHUs and many available agri-environmental measures could be attained.

Study Region Introduction

3

This chapter aims to give a description of the study region in this thesis. First, the general situation of the study region, from the perspectives of physical environment, location and climate, and socio-economic structure, as well as crop production situation, is demonstrated. Second, regarding environmental problems, the soil and water conservation problems and policies in the area are discussed.

3.1 General Situation

3.1.1 Physical Environment, Location and Climate

The study region is a small watershed with the name of "Baishahe"[1], as the carried sediments in the river is eye-seen as white. It is located in the western edge of Zhongtiao Mountain and is within the Loess Plateau area in China. Administratively, it is in Xia county, Yuncheng city[2] in Shanxi province in China. There are several villages in Yaofeng town go through the Baishahe watershed. According to the material from local Water Conservancy Bureau in Xia county (Water

---

[1] The meaning is white sand river.

[2] Here, according to the administrative division in China, the term "Yuncheng city" refers to the whole area of Yuncheng administrative region, including all the urban area and rural area under this administrative region. In this thesis, the referred area of "Yuncheng city" and "Yuncheng" is consistently, i.e. the whole area of the administrative region.

Supplementary Information The online version contains supplementary material available at https://doi.org/10.1007/978-3-658-41340-8_3.

Z. Hao, *An Integrated Modelling Approach to Design Cost-Effective AES for Agricultural Soil Erosion and Water Pollution*, https://doi.org/10.1007/978-3-658-41340-8_3

Conservancy Bureau in Xia county (WCBX) 2008), the Baishahe watershed is located in the suburb of the center of Xia county, with the east longitude scope of $111°01' \sim 111°04'$, the north latitude scope of $34°56' \sim 35°15'$, and the total drainage area of 76.4 km^2.

There is a monitoring station for the water quality of the Baishahe watershed, which is set a bit upstream of the total outlet of the 76.4 km^2 of the drainage area, with the east longitude and northern latitude being $111°16'46.9"$ and $35°4'57.2"$ respectively. Due to the data requirement for this study and the data availability in the monitoring station, the scope of the study region is based on the location of this monitoring station. That is the point decided by the longitude and latitude of this monitoring station is acting as the final outlet of the study area for the Baishahe watershed in this study. The controlled watershed area of this monitoring station for the Baishahe watershed is recorded as 55.9 km^2 according to the material from local Water Conservancy Bureau in Xia county (Water Conservancy Bureau in Xia county (WCBX) 2008). Based on this, with the land-use data in 2015 (Resources and Environmental Science Data Center 2015b), 41.39% of the studied watershed area is rain-fed cropland, 45.66% and 12.95% of it are different types of forest and pasture respectively.

From the perspective of geographical distribution of watershed in China, the Baishahe watershed in Xia county is a small tributary of Sushui River within the Yuncheng city in Shanxi province. Sushui River is one of the first grade tributaries of Yellow River. As showed in Figure 3.1, it is located in the southeast part of the middle reaches of Yellow River. The Yuncheng city has a semi-arid monsoonal climate, with the mean annual temperature of $12.5 \sim 13.5$°C, the mean annual precipitation of ca $500 \sim 600$ mm, and over about $350 \sim 400$ mm of precipitation falling between July and September (Huang et al. 2007). The annual average evapotranspiration in the basin is 1240 mm, which exceeds the average annual rainfall (Li et al. 2015). For getting a general impression for the watershed, please refer to Figure A.7 in Appendix 7 in the Electronic Supplementary Material.

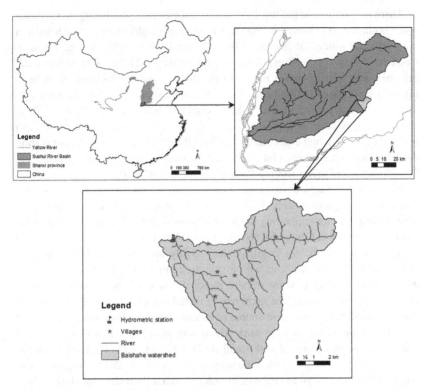

Figure 3.1 Maps showing the location and villages of the study region (Source: Own results with ArcGIS, with data source of Resources and Environmental Science Data Center (2015a))

3.1.2 Socio-Economic Structure

It is important to protect the water quality of the Baishahe watershed. There is a reservoir which was built in the end of the last century and was located in about two kilometers of the downstream area of the monitoring station of the Baishahe watershed. The purpose of the reservoir is to supply drinking water for the citizens who live in the central areas of Xia county and Yuncheng city. The major source of the water in the reservoir is from the Baishahe watershed. The location of reservoir is basically in the boundary of the mountainous area and plain.

There are no firms in the Baishahe watershed area for contributing point source water pollution. As showed in Figure 3.1, there are eight villages which located in the studied watershed (names of the villages: Damiao, Shentouling, Jiandihe, Guojiahe, Houpo, Peipeiling, Jingcao, and Shaling). These eight villages are the ones with administrative record in the county, and the location of them in Figure 3.1 are based on the buildings of their corresponding village committees. There are actually more natural villages scattered in the nearby. Although these natural villages have their special names for the local residents, each of them belongs to the different eight administrative villages respectively. The population density in the watershed is a bit less than the county average level (approximately 266 residents/km^2) (Xia County People's Government 2020).

It is a rural and agricultural area. As it is mountainous, forests account for the majority area (approximately 46%), the second major land use (approximately 41%) is cropland for supporting the local residents' life (Resources and Environmental Science Data Center 2015b). As it is situated in the mountainous area and is in the suburb, residents in villages are mostly all farmers and are relatively poor compared to the residents who live in the central area of the county. Within the Baishahe watershed, there are also minor differences among villages regarding economic well-being. The main reason include cropland quality, location and transport conditions. The main source of income for the village people are small household farming and migrant work. The dominant farming activity are corn and wheat cultivation, combined with small size household level of animal raising in some villages and other sideline farming practices (NLW 2013). Migrant work refers to that farmers go to city area in order to find a short-time job during slack farming season.

3.1.3 Crop Production

Same as most of the situation in China, crop production in the Baishahe watershed is conducted at household level in scattered small patches of cropland. In order to have a general recognition of the crop types in the Baishahe watershed, a statistic analysis for the situation of Yuncheng city from 1952 to 2017 is shown as in Figure 3.2. the Baishahe watershed is located in Yuncheng city. As a hydrological spatial area, it is hard to get the economic statistical data for only the scope of the Baishahe watershed. Although the farming situation of the Baishahe watershed cannot be represented by that of a bigger area which include it, some general characteristics would be suggested. Due to the same crop production market and the similar climate and land conditions between the Baishahe watershed and Yuncheng city, the major crop species are same.

As showed in Figure 3.2, since 2005 the dominated crop types are wheat and corn in Yuncheng, along with the slightly increase of vegetable planting. From 1952 to 2017, along with the increasing trend of area for corn, some other crops' area showed decreasing trends in general, including millet, sorghum, cotton. Tubers and oil crops had relatively high planting area in the middle of the period between 1952 and 2017. Planting area for beans only started since 2000, with a decreasing trend in general. Tobacco had relatively much less planting area compared to others for during the whole period.

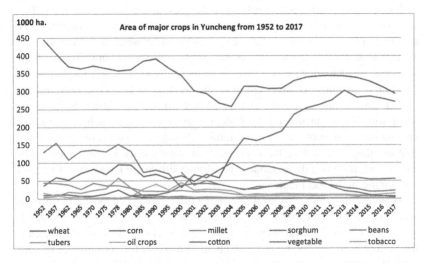

Figure 3.2 Sowing area of major crops in Yuncheng city from 1952 to 2017 (Source: Data from Statistical Yearbook 2015, 2018 and 2019 of Yuncheng city, Shanxi province)

More specified information about the crop production situation in the Baishahe watershed was got from the local farmers and governmental officers (local farmers, personal communication, July, 2018; J.J. Jin, C. Guo, Water Conservancy Bureau, Xia county, personal communication, January, 2018), which is described as follows. First, as consistent with the situation in Figure 3.2, the major crops are corn and wheat, with minor area for vegetables (mainly tomato), medicinal plants, oil crops (oilseed rape), cash trees (mainly pepper), and others. In recent years, more and more farmers in the watershed started to plant cash trees of pepper, medicinal plants and vegetables, which induces the decrease of area for wheat and corn. This is actually a general situation in Yuncheng for reducing the planting of wheat and corn and preferring more economic plants on cropland, as showed in Figure 3.2. For some patches of cropland close to the river, the cultivation of winter wheat and corn in succession in one year is operated. That is after the harvest of winter wheat (usually in the beginning of June) the corn is planted immediately and is harvested usually in September. This is because of the semi-arid climate and the shortcoming of efficient irrigation system in the watershed, thus only the cropland near the river have conditions to be manually irrigated when needed. On contrast, most patches of cropland in the area are upland terraces, which are usually planted only with winter wheat during the whole year, due to wheat's feature of drought resistance.

Second, almost all cropland patches in the area are terraces from history. This is because the natural geographical terrain condition in the area. It is mountainous, therefore if farmers want to plant crops and harvest well, they have to build terraces to make the land to be farmable. These terraces are built along with the mountain contour, which makes them to be mostly shaped as very long along the direction of mountain contour and relatively short along the vertical direction of contour line. Farmers have the knowledge and experience that doing the tillage along with the mountain contour (longer side of cropland patch) is labor and time saving because of less times for turning during the tillage activities, and that contour tillage can reduce the damage of cropland caused by rain, such as the loss of soil. Based on these, terrace and contour tillage are the historical features for farming in the local region.

Third, the mechanization level of farming in the watershed increased since the beginning of the new century. One reason is that along with the economic development of China the local farmers in the villages also have increasing income, which implies the increasing affordability of farmers for the consumption of agricultural machines. Another reason is the various agricultural support policies in China, which help farmers to get more and more subsidies and supports for improving the situation of poverty. For example, for buying no-till planters farmers can get a certain percentage of reimbursement as subsidy from government. In the past decade, most of the terraces are ploughed by walking tractors, and most of households in the villages have had an agricultural tractor. More recently, combined-harvesters for wheat and corn are adopted in the local area. However, many terraces in the Baishahe watershed are in upland and do not have good road conditions for big combined-harvesters to operate. Along with these development, the new form for adopting modern agricultural machinery on farmland in China and the study region area is farming trusteeship. It means that, taking the example of harvester, farmers hire a harvester together with a professional operator to finish the harvest work for a special crop in a certain area of cropland and then pay the corresponding agency according to various criteria, such as the crop type and the area of cropland being performed.

3.2 Soil and Water Conservation in the Watershed

3.2.1 Relevant Problems

The cultivation practices on the cropland are recognized as the major contributor to water quality degradation throughout China (Mateo-Sagasta et al. 2013). This is the especially the situation in the Baishahe watershed. According to a report from the local Environmental Protection Bureau in Xia county (Environmental Protection Bureau in Xia county (EPBX) 2012), the Baishahe reservoir and the nearby area was fixed as the first one-level water source protected area. In the report, the required protective measures for the water quality in the Baishahe reservoir were descried, which emphasized two points related to the upstream farming activities. One is to reduce the sediments in the reservoir caused by soil erosion from upstream, the other is to reduce the fertilizer application on cropland in upstream area.

Regarding soil erosion, the natural reason include the large land slope as in mountainous area and the special soil texture there. However, anthropogenic causes are significant, which mainly refer to the cultivation activities on cropland, including tillage practices and bared cropland in some periods of the year. As for nutrient pollution in the Baishahe reservoir, reducing cropland fertilizer is explicitly recommended in the report (Environmental Protection Bureau in Xia county (EPBX) 2012). As the Baishahe reservoir servers as the drinking water source, and firms acting as the point source pollution are prohibited in the nearby area. Therefore, the sources for nutrient pollution are cropland fertilizer application, livestock raising, and resident living. Among these, the contribution of the latter two are quite minor. The major problem is fertilizer application on cropland. According to the statistics in the report (Environmental Protection Bureau in Xia county (EPBX) 2012), there are about 80% and 85% of cropland in Xia county which applied pesticide and chemical fertilizer respectively, with the average implemented amounts of 2.4 kg/ha and 763.5 kg/ha respectively.

3.2.2　Protection Policies

Government have promoted and implemented many polices and measures for the soil and water conservation in the Baishahe watershed area. The government in Xia county has been doing the work of testing effective conservation measures under the local conditions in recent years. According to the work summary report from Agricultural Bureau in Xia county in 2015 (Agricultural Bureau in Xia county (ABX) 2015), many protection measures aimed on cropland are tested and promoted. These measures include "soil testing and fertilizer recommendation", "increasing organic fertilize", "residue return", "reduced tillage and no-till", "cash crop hedgerows", and others (the main recommended measures are summarized in Table A.4 in Appendix 1 in the Electronic Supplementary Material). "Soil testing and fertilizer recommendation refers" to the local government doing experiments for obtaining the appropriate fertilizer formulas and corresponding application amounts regarding the crop yield expected. With this, the government has given recommendations for different crop types, like the ones for winter wheat and corn as described in Table 3.1. "Increasing organic fertilizer" refers to encouraging farmers to apply more organic fertilizer on cropland (e.g. green manure, human and animal waste, commercial organic fertilizer) instead of chemical fertilizer. "Residue return" indicates to leave the residue of crop after harvest (e.g. straws and roots of corn and wheat) on the cropland for turning them into organic manure, instead of collecting the residue and burning them. "Reduced tillage and no-till" refers to reducing the tillage activity on cropland as much as possible. "Cash crop hedgerows" means that planting economic trees or shrubs in a set or series of single or multiple rows, with other crops planted in the alleys between the rows of woody plants.

Table 3.1 Recommendation for fertilizer application in Xia county

		Amount (unit: kg/ha)		
Wheat on dryland	expected yield	> 4500	3000~4500	< 3000
	fertilizer formula: N-P_2O_5-K_2O with 15-10-5	900	750	600
Corn	expected yield	> 9000	6750~9000	6000~6750
	fertilizer formula: N-P_2O_5-K_2O with 25-11-4	600 (+525 urea later)	450 (+450 urea later)	375 (+375 urea later)

Source: Content adapted from Agricultural Bureau in Xia county (ABX) (2015)

However, although highly promoted by the local government, there are no compensation payments for farmers to implement these measures. These recommended measures are barely adopted in the Baishahe watershed, except some activities being similar like the measure of "Cash crop hedgerows". In recent years, a small amount of farmers (about 30%) in the Baishahe watershed started to plant on their cropland cash trees voluntarily, especially pepper trees (local farmers, personal communication, July, 2018). It is mainly driven by that the net economic benefits from planting pepper trees are more than the traditional crops for some farmers. This is helpful for preventing soil erosion and nutrient excess in the area, as land with permanent trees and orchards are significantly better than arable land for retaining soil and nutrients (Meng et al. 2001).

The most influential policy and conservation project in the study watershed is the national Grain-to-Green program in China. It required all the cropland with the slope degree of more than 25 to convert to all kinds of forests or grassland since 2000. Under this program, a big percentage of arable land in the Baishahe watershed has been planted with all kinds of trees by now. For example, among the eight villages in the Baishahe watershed, the villages of Shentouling, Guojiahe, and Shaling have participated in the Grain-to-Green program with the arable land area of 28.4 ha, 20 ha, and 65.9 ha respectively, which account for their original arable land of 20%, 25% and 54.9% respectively (Yunchengshi Nonglianwang 2013).

Overview of Integrated Hydro-Economic Modeling Approach

4

This chapter gives the overall structure of the integrated hydro-economic modelling approach in this study, and shows the important details for cooperating and integrating all involved interdisciplinary components in the integrated modelling procedure. First, the technical routes in the developed hydro-economic modeling procedure are introduced. Second, the means to link and coordinate the interdisciplinary components in the integrated modelling procedure is described.

4.1 Overall Structure of Hydro-Economic Modelling Procedure

An integrated hydro-economic modelling procedure is applied to achieve the aim of this study. A graphical framework to illustrate this procedure is shown in Figure 4.1, where different components are connected with technical routes. These components are interdisciplinary, with each of them refers to a specific corresponding function and task. The interdisciplinary knowledge involved include engineering and agronomy for the technical details of agri-environmental measures, hydrology and ecology for assessing the mitigation effects of the measures, economics for opportunity cost assessment of the measures, and numerical analogy of computer for the simulation of farmers and optimization work. The challenge is that due to the interdisciplinary tasks each component needs to be carried out separately, but at the same time these interdisciplinary tasks have to be linked into a systematic framework in a consistent way in order to achieve the final study aim.

© The Author(s), under exclusive license to Springer Fachmedien Wiesbaden GmbH, part of Springer Nature 2023
Z. Hao, *An Integrated Modelling Approach to Design Cost-Effective AES for Agricultural Soil Erosion and Water Pollution*,
https://doi.org/10.1007/978-3-658-41340-8_4

Based on this, the framework with hydro-economic modelling procedure is developed and illustrated in Figure 4.1. The logic for this integrated framework is mainly inspired by Wätzold et al. (2016), which sets the target of obtaining cost-effective AES designs for biodiversity conservation with spatiotemporal heterogeneity to be emphasized. However, this study focuses on another specific environmental problem of agricultural NPS water pollution from cropland, with the spatial heterogeneity to be the key for achieving the cost-effectiveness of AES designs. The technical routes of the integrated hydro-economic modelling procedure in this study are described as follows.

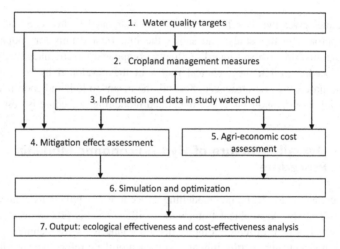

Figure 4.1 Overview of hydro-economic modelling procedure (Source: Own analysis and drawing)

In this procedure, the first step is the identification of the specified targets of water quality (Figure 4.1, box 1). Based on the situation of China and the study region, three indices of NPS water pollutants are targeted, which are sediments, nitrogen and phosphorus. The target is to mitigate the amount of loads of these three pollutants in waterbodies caused from cropland cultivation. On the basis of the target identification, cropland management measures that have the corresponding potential to mitigate the loads of sediments, nitrogen and phosphorus are needed to be identified as the original measure inventory for this study (Figure 4.1, box 2). Each of these identified measures should be functional for at least one of the pollutants, while one individual measure can also be useful for two or more pollutants mitigation. Besides, the identification of measures needs to follow some principles in the integrated study, such as being suitable to the study region conditions, cooperative with later components of mitigation effect assessment and cost estimation, and adaptive for AES projects. Data and information in the study region (Figure 4.1, box 3) is significant for the relevant components of measure identification, mitigation effect assessment and cost estimation (box 2, box 4, and box 5 in Figure 4.1 respectively). Both quantitative and qualitative as well as both primary and secondary data are required, with data source including local government reports, questionnaire surveys, personal communications, and scientific database websites (mainly for SWAT model setup, please refer to Chapter 6). The acquired data are pre-processed differently according to their corresponding usages for different components in the integrated procedure.

Based on the first three steps, mitigation effect assessment and agri-economic cost estimation (box 4 and box 5 in Figure 4.1) could be processed side by side. The aim of these two components is to obtain the quantitative intermediate results of this study in both the ecological and economic perspectives, regarding scenarios of implementing each identified measure in each heterogeneous spatial unit. Therefore, the study region would be divided into many SHUs, where each measure is assessed for its quantitative mitigation effect and opportunity cost. Regarding mitigation effect assessment (Figure 4.1, box 4), the eco-hydrological model of SWAT is adopted in this study for conducting the simulation. The SWAT model is first built for the study region with the data from box 3 in the procedure, and then it divides the study region into SHUs according to hydrological principles. Afterwards, each identified measure from box 2 in the procedure is simulated in SWAT model for getting its mitigation effects regarding the targeted pollutants in box 1. Meanwhile, for agri-economic costs assessment (Figure 4.1, box 5), first the cost categories for each identified measure from box 2 and their corresponding calculation methods and formulas are analyzed and established; then the mathematic calculations of cost estimation for each measure are processed based on the data from box 3 in the procedure in Figure 4.1.

Based on the above, the intermediate results of mitigation effects and agri-economic costs of measures from box 4 and box 5 respectively are combined together for the procedure of simulation and optimization (Figure 4.1, box 6). Simulation is to mimic farmers' selection behaviors when in AES programs farmers are offered with measures along with their payment and costs in SHUs. Optimization is to decide the sets of payments of measures which could lead to the maximum total mitigation effects, based on farmers' selection behaviors and mitigation effects of measures in SHUs, under the limitation of budget level of AES programs. The heuristic optimization method of simulated annealing is applied in this process. Due to the complexity, the processes for simulation and optimization are synthesized into an optimization modelling procedure developed by a computer scientist. The combined data for mitigation effects and agri-economic costs of measures in SHUs could be input into the optimization modelling procedure, which operates the simulation and optimization processes with options for budget levels and available measures. The procedure then could result in cost-effective AES designs with all kinds of detailed information, such as total amount of given budget, final mitigation effects in quantification, the offered measures to farmers and their corresponding payments respectively. At the end, with obtained results of cost-effective AES designs, the further analysis for them in terms of ecological effectiveness and cost-effectiveness is conducted and concluded (Figure 4.1, box 7).

4.2   Coordination and Cooperation of Components

4.2.1   Inter-Limitations Between Components

One of the challenges in the integrated hydro-economic modelling approach is to identify appropriate tools and carriers for each component, especially the ones from box 2 to box 5 in the procedure in Figure 4.1. The selection of study region, the identification of measures, the identification of eco-hydrological model, and the method for cost estimation are not independent, while the interconnection between each of them are important for making the befitting choice for each of them so that the they can be integrated together. The analysis for their interconnection between each other is needed for obtaining some clues of the limitations and requirements for each single part. The aim of it is to summarize a list of limitations for each individual part, under which the task of each part can be conducted separately and at the same time the task of each part can also be carried out in a way of coordination and cooperation within the integrated framework. The analysis for their interconnections and inter-limitations are illustrated in Figure 4.2, with the explanation in below.

As showed in Figure 4.2-A, it is not appropriate to just select and identify a tool or carrier for each part of the component in this integrated interdisciplinary study. To make sure the cooperation between each component, the influence of one component to the other is profoundly considered. In Figure 4.2-A, four interconnected components are considered, with the possible limitations for the selection and identification of exact tool or carrier in one component being analyzed one by one by the influence of other three components.

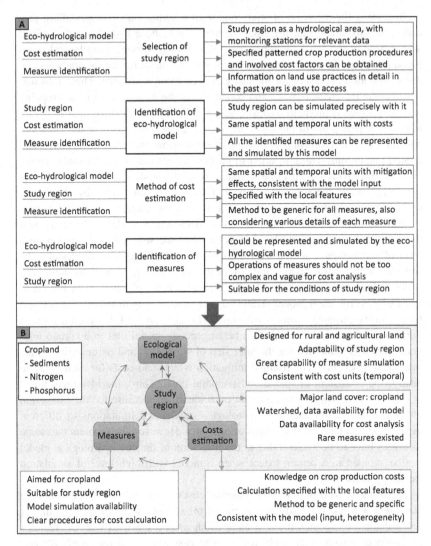

Figure 4.2 Coordination and cooperation of interdisciplinary components (Source: Own analysis and drawing)

Regarding the selection of study region in a hydro-economic modelling procedure in this study, first the study region need to be a hydrologically defined area instead of the usual administrative scope in order for eco-hydrological simulation of this study region. Second, for identifying measures for the study region, the information necessary for the analysis need to be accessible. These information include the characteristics and conditions of the area, such as main cropping activities which is prone to lead to loads of pollution; the detailed land use practices in the past years in the study area acting as the business as usual situation; and the possible promising promoting methods from the local government and related institutions. Third, in order to do the cost estimation in the study region, clarified information need to be available from the study region on the features of local farming, the patterned crop production procedures and also the details involving spatial heterogeneity, as well as the various costs, yields, and others.

For the identification of the proper eco-hydrological model in the study, first the identified model should be able to simulate the selected study region on a level of stable representation for the watershed, which requires the model to be suitable for the characteristics and size of the study region. Second, the simulated quantified intermediate results of mitigation effects of measures from this identified eco-hydrological model should have the same units in terms of space and time with the estimated costs of measures. This is because of that each measure's mitigation effects and costs in each heterogeneous spatial unit in a certain temporal period is combined for the later step of simulation and optimization in the integrated procedure. The resulted mitigation effects and costs for each measure need to be consistent in spatial and temporal units. Third, the identified model is required to be capable to represent each identified measure for the mitigation effect simulation. As each eco-hydrological model has limitations in different aspects for simulating various types of measures. However, the problem that some measures have to be avoided due to the limitation of the eco-hydrological model can be avoided in a certain extent when a model with high level of simulation capability is identified.

As to the method of agri-environmental cost estimation, first it needs to keep consistent with the eco-hydrological model for spatial and temporal units during cost calculation. Second, the cost analysis should be specified with the features of the local situations of study region. The cost estimation for cropland management measures is mainly to obtain the foregone profits of farmers when they change from their traditional practices to the requirements of identified measures. Therefore, during the cost analysis many processes need to be considered, including the general knowledge on crop production costs analysis, the local traditional farming procedures for each of the major crop types, and the changed practices and procedures when implementing identified measures in each of the different crops. Third, for developing the cost calculation formulas for measures, it should be generic for the application of all measures and similar studies, as well as be flexible and specific for being able to consider and do the calculation of the various details of different measures.

The identification of measures is also influenced from other components, which basically forms some of the major principles for identifying appropriate measures in this study. First, all the identified measures should be aimed at the specific problems in the study region and be suitable to be applied in the area based on the characteristics and conditions of the study region. Second, all identified measures must be able to input into the eco-hydrological model for simulation. That is there are proper ways to represent each of these measures through the adjustment of the model parameters with reasonable quantified results being obtained. Third, considering to do the cost estimation of the identified measures, the specific practical operations of each measure need to be analyzed step by step and to make sure that each step of the operations can be specified in terms of cost calculation. Therefore, measures with vague description for their practical procedures and the ones with difficult and complex operation details might not be suitable for voluntary farmer application and for rational cost estimation.

The interconnection of these components could be further concluded as illustrated in Figure 4.2-B. The study region is in the core of the interrelationship for being the basis to connect all components. Each of the components has impacts to others and gets reversed feedbacks from others. The water quality targets in this study are to mitigate loads of pollutants of sediments, nitrogen and phosphorus from cropland. Therefore, the study region should be mainly covered with the land use of cropland; the identified measures need to be aimed and suitable for the location on cropland; the eco-hydrological model should be the one designed for the simulation of rural and agricultural land, especially cropland; and for cost estimation, the knowledge on the analysis of framing procedures on cropland and crop production costs should be first demanded. Combined this with the interconnection of the related components illustrated in Figure 4.2-A and explained in above, the basic clues of requirements and inter-limitations for each component, with the aim of coordination and cooperation in an integrated study, could be summarized as listed and showed in Figure 4.2-B.

With the analyzed results of the listed limitations for each component in Figure 4.2-B, the selection and identification of tools and carriers for the task of each component can be conducted, with the aim of cooperating the components in an integrated procedure. However, it is not that following the listed inter-limitations the task of each component can be carried out alone and step by step. The interconnections exist in many details from various perspectives, and the components interact each other as in the way of forming a circle instead of orderly (Figure 4.2-B). Therefore, the appropriate selection and identification of tools and carriers for components in the integrated approach should always be turned out as a set of results for all components but not as separated decisions for each single part.

Under this logic, the study region for this research has been described in the previous section of Chapter 3. The detailed information for each of the other components is demonstrated in the following chapters respectively.

4.2.2   Spatial and Temporal Consistency

Some operational and technical difficulties might appear in the process of the integrated interdisciplinary studies. In integrated hydro-economic modelling, this aspect is mainly reflected on the spatial and temporal consistency between eco-hydrological model and economic cost estimation (Brouwer and Hofkes 2008). From the perspective of spatial units, as eco-hydrological models are mainly designed based on the hydrological principles, and the geographical units in all kinds of levels in the model simulation mainly refer to hydrological units as a whole, such as basins, watersheds and sub-watersheds. However, for cost estimation in the economic field, administrative boundaries of regions, like state, province, county and town, are the commonly adopted spatial units for analysis and statistics. Regarding temporal units, in eco-hydrological models the units often refer to days, months and seasons, while in economic analysis the time scales are usually longer than that, like years.

In this study, for each identified measure the mitigation effect simulated from eco-hydrological model and the opportunity cost estimated with economic analysis need to be consistent for having the same spatial units and time scales. The eco-hydrological model of SWAT is adopted in this study, with which the study region is shaped as a hydrological basin. SWAT could divide the study region into many sub-watersheds and could further divide the sub-watershed into many spatially heterogeneous HRUs (hydrological response units). In this thesis, the study region of the Baishahe watershed is divided by the SWAT model into 83 sub-watersheds (50 sub-watersheds are covered with cropland), and for each sub-watershed there is no further division for more HRUs (therefore one sub-watershed is acting as one HRU). These divided sub-watersheds are geographically acting as the SHUs for this study. The geographical units of the 50 sub-watersheds with land cover being mainly cropland are acting as the focused

SHUs in this study for assessing the mitigation effect of measures and estimating the economic costs of measures.

With this, the underlying assumption is that the resulted SHUs are acting as the different farms in reality with heterogeneity. These SHUs geographically result from the divided sub-watersheds in SWAT, which in reality are not the usual way to distinguish farms. The thesis assumes that in each of these heterogeneous farms there is a representative farm for making decisions for this farm, and each farm represents a homogeneous unit with the same conditions and characteristics regarding land use, soil, farming activities, etc.

Regarding the time scale, the temporal unit of year is adopted in this study. For the mitigation effects of measures simulated by SWAT model, yearly results are oriented. With the aim of cost-effective AES designs, this study sets the assumption that the period of AES is five years from 2018 to 2022. Due to the climate data limitation for SWAT model from 2018 to 2022, the average annual results of mitigation effects of measures from 2013 to 2016 resulted from the SWAT model is assumed to be the yearly average mitigation effects of measures in the AES design lifetime in this study, i.e. from 2018 to 2020. To keep consistent with the time scales, the economic costs of measures need also to be calculated as average annual results from 2018 to 2020 in the set AES design lifetime. To do this, the original data for cost analysis are collected mainly through questionnaire surveys in 2018, based on which the yearly average costs of measures are calculated according to the related principles of temporal discounting. With these operations, the intermediate results of both the mitigation effects of measures from SWAT model and the costs of measures from agri-economic analysis share the same temporal unit of yearly average outcomes from 2018 to 2020.

According to the above description, it is noticeable that along with the addressing for the consistency of spatial and temporal units in the integrated interdisciplinary hydro-economic modeling approach some underlying assumptions have to be made and in the process data availability is difficult. Regarding the consistency of spatial scales, in this study the geographically uniform spatial units are based on the ones formed from hydrological rules. These spatial units are shaped with their boundaries in hydrology principles. For doing the cost estimation these spatial units need to be assumed as heterogeneous farms. This brings the difficulty that the information of economic statistics which is usually carried out in the scope of administrative regions can not be applicable in this study. The heterogeneity of cost estimation of measures in this study is mainly considered from the perspective of the locations of each spatial unit and their corresponding data in the map. For the consistency of temporal scales, yearly average results in the period of 2018 to 2022 are expected for mitigation effects and costs of measures. Assumptions are made as described before for solving the problem of data difficulties. The detailed information in this process for the assumptions and the corresponding operational techniques are demonstrated in Chapter 6 and Chapter 7.

Identification of Mitigation Measures 5

In this chapter the logic and process on how to identify the appropriate mitigation measures for this study is described. First, the potential measures, the best management practices, are introduced with the concept, development and classification of them. Second, the specific criteria for identifying appropriates measures for this study are deeply demonstrated. Third, based on the identified criteria for measure selection, the final selected measures for this study are listed with each of them being explained in detail.

5.1 Best Management Practices

5.1.1 Concept

The increased awareness of the demand to improve water quality resulted in the concept of BMPs (best management practices). BMPs mainly refer to land use measures which intended to provide an on-the-ground practical solution to NPS pollutions from all sources and sectors (D'Arcy and Frost 2001; Field and Tafuri 2006). There are lists of defined measures and practices worldwide under the concept of BMPs for different sectors, including forestry, agriculture, urban development and others. The goal of these practices is either to maintain and replicate the pre-development conditions of the watershed ecosystem as close as

Supplementary Information The online version contains supplementary material available at https://doi.org/10.1007/978-3-658-41340-8_5.

Z. Hao, *An Integrated Modelling Approach to Design Cost-Effective AES for Agricultural Soil Erosion and Water Pollution*,
https://doi.org/10.1007/978-3-658-41340-8_5

possible, or to reduce the environmental impacts to an acceptable level (Paule-Mercado et al. 2017; Field and Tafuri 2006).

As said by Ritter and Shirmohammadi (2000) and Evans et al. (2003), there is unfortunately no universally accepted definition of BMPs. The Soil and Water Conservation Society (SWCS) defines a BMP as "a practice or usually a combination of practices that are determined by a state or a designated area-wide planning agency to be the most effective and practical (including technological, economic, and institutional considerations) means of controlling point and non-point source pollutants at levels compatible with environmental quality goals" (Ritter and Shirmohammadi 2000, p. 259). An another generally alternative definition is described as "BMPs are methods and practices or combination of practices for preventing or reducing nonpoint source pollution to level compatible with water quality goals" (Ritter and Shirmohammadi 2000, p. 260). However, both definitions state the consistent purpose of BMPs, which is to reduce pollution levels to achieve water quality goals. When referring to rural areas, BMPs are often called as conservation practices or agricultural and silvicultural BMPs (Evans et al. 2003).

According to Boyd (2003, p. 106), BMPs are generally considered to be "the best available and practical means of preventing a particular environmental impact while still allowing production to be conducted in an economically efficient manner". However, the author also emphasized that the word "best" is not intended to imply that a particular BMP will always be the best practice, in contrast, the "best" practices must be selected based on site characteristics, and as technology advances, BMPs must be revised to reflect new knowledge. Because of this, some researchers do not like the term BMP and insist on referring to "good management practices" or "better management practices", while as demonstrated by Boyd (2003), the term BMP is still acceptable terminology in environmental management.

There is usually a very wide range of BMPs that could potentially be employed for a region, which could result in correspondingly a wide range of inherent pollutant reduction efficiencies and associated costs. As the contribution of pollutants is spatially disproportionate in a watershed (Maringanti et al. 2009), the potential effectiveness of BMPs is also site specific. The ecological effects and economic costs of BMPs vary due to many specific factors, such as varying design methods and implementations, frequency of maintenance, and placement in the watershed also plays a vital role (Giri et al. 2012). When considering options of BMPs in a certain special region, under some specific environmental goals and budget limitations, it is useful and necessary to have the information of both the ecological effects and economic costs of each potential BMP (Duffy 2014).

The cost-effective analysis of BMPs regarding spatial heterogeneity is increasingly demanded and has been attracting increasingly attention. Previous research generally focuses more on the ecological effects of BMPs, among which there are mainly two ways for the assessment of the ecological effects of BMPs. A direct way is to measure the changed pollution loads from all kinds of field in a watershed through actual implementations, while this is kind of impractical considering the resource and time consumption. On the contrary, watershed eco-hydrological modelling is preferred for BMPs implementation strategies, as it is considered to be efficient for providing information which is needed for evaluating pollution loads (Giri et al. 2012).

5.1.2 History and Development

Research on the control and management for agricultural NPS pollution first started in 1970s in America, developed greatly in 1990s and is now represented mostly by the Best Management Practices (BMPs) developed in America (Jiang et al. 2006). BMPs represent a management response to water quality problems. It is first proposed based on the concept of rationalization of land-use by the Federal Water Pollution Control Act in 1972 in the USA, and is initially designed for the purpose of controlling soil erosion (Zhuang et al. 2016). In the 1980s programs with BMPs were increasingly designed mainly for addressing pollution from wet-weather flow, focused on controlling runoff and water quality degradation (Field and Tafuri 2006).

BMPs are inherently pollution prevention practices, which have been recommended as an effective tool used widely to reduce NPS water pollutions. United States, United Kingdom, and other European countries were the first to implement BMPs broadly since 1970s mainly for the agricultural NPS water pollution problems, and have obtained successful experiences (Jiang et al. 2006; Zhuang et al. 2016).

Among the national BMP programs, the representative ones which promoted the broader implementations of BMPs include the Conservation Effect Assessment Project in the USA (Simon and Klimetz 2008), the Watershed Evaluation of Beneficial Management Practices in Canada (Yang et al. 2007), and the Systems Approach to Environmentally Acceptable Farming in the European Union (Turpin et al. 2005). In the meantime, over the last a few decades, BMPs have been applied in an accelerated tempo worldwide, with continuously new practices being put forward (Zhuang et al. 2016). Research on BMP starts later in China compared to other the counties mentioned, without systematic BMPs based on the specific land characteristics being built (Tang 2010). For achieving the aim of preventing soil erosion and nutrient pollution from the cropland to waterbodies, many tested and implemented measures in China are learned from the above mentioned countries.

5.1.3 Classification

BMPs can be classified into many different kinds of categories from different perspectives. Here three kinds of categories are described, with BMPs being classified from the perspectives of the location of BMPs from source areas to waterbodies, the sector that BMPs belongs, and the duration of the lifetime of BMPs.

(I) Location
From the perspective of the location of NPS water pollution controlling practices, there are mainly three kinds being classified. These refer to source control techniques on the original sites of pollutants happened, process control techniques and transport interception located between source sites and waterbodies, and terminal treatment (Shen et al. 2014a; Sun et al. 2013). Management practices with source control techniques means that the practices are designed for the locations where the NPS pollutions are originally happened, such as conservation tillage on cropland. Management practices with process control techniques indicate that the practices are designed for the function of reducing the loads of NPS pollutants in the process of transport between the original areas of pollution generation and the final waterbodies, such as riparian vegetation buffers. Management practices with the techniques of terminal treatment represent the practices that are focused to deal with the already polluted water in rivers, lakes and other waterbodies, such as returning straws in the waterbodies to the field.

For BMPs, especially for cropland BMPs, source control techniques and transport interception are the focus, which are either preventing the runoff from becoming polluted (source control) or to treat the polluted runoff before it reaches a water system (transport interception) (Wyoming Department of Environmental Quality 2013).

(II) Sector
From the perspective of sectors, BMPs can be divided as the ones designed for urban areas and the ones designed for rural and agricultural areas. There are detailed information and study reviews which are aimed for the comparison and category of urban BMPs and rural BMPs, such as the research by D'Arcy and

Frost (2001) and Liu et al. (2017). However, except the classification of BMPs for urban area and rural area, more specific divisions of BMPs from the perspective of sectors are shown, such as in terms of categories of forest, cropland, and marinas. Among these specific categories, the majority of studies focuses on BMPs designed for forest and cropland.

Regarding this study, the focus refers to agricultural BMPs. According to D'Arcy and Frost (2001), there are three classes of NPS agricultural pollutants, i.e. nutrients (nitrogen and phosphorus) from fertilizer and animal manures, pesticides, and suspended solids from soil erosion. There are different agricultural BMPs designed either for the individual aim of nutrients, pesticides, and sediments, or for the different kinds of combination aims of them.

(III) Duration
From another perspective of the duration of lifetime of BMPs, there are two kinds of them being categorized, which are operational practices[1] and structural practices[2] (Campbell et al. 2005). These are the major categories normally being demonstrated for BMPs in the related studies. The difference exists in the implemented process and the duration of the conservation practices themselves. According to Duffy (2014), operational conservation practices are short-run practices that can be implemented on a year-by-year basis; while permanent conservation practices are long-term practices that will remain in place until they are removed or altered.

In agriculture, operational practices mainly refer to the tillage and other cropland management activities based on the knowledge and experience of farmers, and thus they can usually be implemented by farmers directly and according to their annual decisions (Potter et al. 2009). On the contrary, structural practices require more than annual management decisions. These practices are considered to have permanent or long-term effect with only one-time change or implementation, which usually require engineering designs, surveying and even contracting with a professional vendor. Generally, on agricultural land operational BMPs are applied for the aim of source control, while structural BMPs usually function in terms of transport interception. The differences between these two kinds of BMPs are shown in Table 5.1 with some example measures.

However, sometimes it is hard to distinguish one BMP for being the category of operational practices or structural practices, as said by Duffy (2014), some management practices may fall under both of these two categories depending on the specific circumstances.

---

[1] Also being termed as cultural practices in the literature.

[2] Also being termed as permanent practices in the literature.

Table 5.1 Difference between operational and structural practices

Category	Description	Examples
operational practices	• short-term practices • implemented on a year-by-year basis • involve changes in farm management • result in changes in annual decision-making	• nutrient management • conservation tillage • cover crops • residue management
structural practices	• long-term practices • require engineering design • remain in place until removed/alter • often control surface runoff	• terraces • field borders • grassed waterways • riparian buffer strips

Source: Content adapted from Duffy (2014), D'Arcy and Frost (2001), Wyoming Department of Environmental Quality (2013)

Based on the above, a not exhaustive list of commonly used BMPs in agriculture (mostly on cropland) is presented in Table A.1 in Appendix 1 in the Electronic Supplementary Material, with Table A.2 and Table A.3 in Appendix 1 in the Electronic Supplementary Material for classifying the BMPs based on these different categories[3]. These listed BMPs are mainly acted as the inventory in this study for the identification of appropriate land use measures.

5.2   Identification Criteria

5.2.1   AES Design Suitability

In this study, the research aim is to obtain cost-effective AES designs with the application of proper land use management measures. It assumes that farmers select and implement the provided measures voluntarily according to each measure's corresponding cost and payment. This implies that the operational practices which are implemented on a year-by-year basis instead of structural practices which needed professional engineering design will be the focus in the process of measure identification for this study. As operational practices are more flexible to be adopted and are much easier for farmers to go back to their traditional farming ways if they needed when the AES programs are over. Compared to

---

[3] For more information on comprehensive list of BMPs regarding all kinds of land uses of cropland, forest and livestock raising, one can refer to NRCS 2019.

structural measures, operational measures will be more preferred by farmers regarding acceptance.

Meanwhile, with AES programs it is expected that farmers have the ability to implement the special measures on their own. Therefore, implicitly, it is better to offer farmers with BMPs that are easy to implement from farmers' perspective, and then they can select these BMPs without worries for the operation complexity. Although the standard of easiness is hard to define here, the ones with obvious complexity and high professional requirement, such as many of the structural practices in Table A.3 in Appendix 1 in the Electronic Supplementary Material, which are not rational to be considered. However, it is not that all structural measures will be excluded. As demonstrated by Campbell et al. (2005), for most situations a suite of BMPs is available, while the combinations of good operational BMPs with provision of one or more structural BMPs are not unusual.

One important point for this study is that the basic thing is that the measures should be prepared for farmers only on the location of cropland. For participating in AES programs, farmers need to have the land use right to change the traditional land-use activities to the land management measures they selected voluntarily. In China, because of the special institution for land uses rights[4], the area where farmers have land use right is only the cropland, but not include the contiguous areas beyond cropland and between the rivers, where lots of BMPs are targeted to locate and are generally effective for acting the function of intercepting the transport of pollutants. Based on this, all the selected measures for this study need to be designed and aimed only for the cropland.

5.2.2 Study Region Suitability

Although there are summarized lists with commonly applied cropland BMPs, such as the ones described in Appendix 1 (Table A.1, Table A.2, Table A.3) in the Electronic Supplementary Material, not all of these measures are suitable for the study region in this study. The suitable ones need to be selected based on the information of the conditions of the study region, as demonstrated in Chapter 3.

In the Baishahe watershed, some land use management measures which are generally recommended for being effective are already implemented by local farmers, including terrace, contour farming, and deep ploughing. These measures have been adopted in the study region mainly due to other reasons instead of AES

---

[4] For detailed information, please refer to Dean and Damm-Luhr 2010.

programs[5]. The already existed measures in the study region, although without payment, like terrace and contour farming, are not considered.

Besides, based on the report for cropland improvement in Xia County from the local Agricultural Bureau (Agricultural Bureau in Xia county (ABX) 2015), many conservation measures for the aim of improving crop yield and cropland ecosystem are recommended but still in the stage of testing. These soil and water conservation measures which aim specially for the pollutant targets of this study (i.e. reduction of sediment yield, nitrogen and phosphorus) are summarized in Table A.4 in Appendix 1 in the Electronic Supplementary Material, which are basically the same or similar as some BMPs included in the initially summarized lists (in Table A.1, Table A.2 and Table A.3 in Appendix 1 in the Electronic Supplementary Material). Although it is a general recommendation for a bigger area in the whole county instead of the area of the study watershed, these measures could act as a major reference for the selection of BMPs for this study.

5.2.3 Applicability for Mitigation Effect Assessment

For assessing the mitigation effects of identified measures, the eco-hydrological model of SWAT is adopted in this study. The identified measures need to be sure that each of them can be represented and simulated by SWAT for getting the quantified mitigation effects. Generally, the SWAT model is capable of simulating a wide variety of land use management measures (Xie et al. 2015). Some researchers even concluded that among the similar eco-hydrological models for agricultural area the SWAT model shows the simulation ability in terms of the greatest number of land use measures (Arabi et al. 2008; Kalin and Hantush 2003).

Compared to the simulation of operational measures with eco-hydrological models, the simulation of structural measures are different and not straightforward. As structural measures usually are implemented in some special positions of the land and occupy relatively smaller area. In eco-hydrological model, it can only represent one kind of crop or plant as the land cover in an individual homogeneous spatial unit under a period of time, but cannot represent the growing of two or more kinds of plants in one spatial unit simultaneously. Due to this, there are two means to represent the structural measures.

One way is to adjust the values of the relevant parameters in the model which are designed to represent the functions of measures (Arabi et al. 2008). With

---

[5] The reasons are analyzed in Chapter 3.

this way, the model cannot tell the detailed information (e.g. exact plant species, specific location) of the structural measures. Therefore, the information for cost analysis of the measure is vague. Besides, the adjustment of the values of parameters could be subjective. The other way is to distinguish the exact location of the structural measure and separate the area out as the new spatial unit, where the features of the structural measure can be configured like that for operational measures. However, this way is not practical for the study region in this thesis due to the low resolution of data for the SWAT model. Based on this, the majority of structural measures are not preferred in the measure identification process for this study.

Another problem with eco-hydrological model for measure simulation is the data availability. As described in the definition of each practice in Appendix 1 in the Electronic Supplementary Material, for measures related to the wind (erosion), the detailed history information refers to the direction and speed of wind in each heterogeneous spatial unit is necessary for the appropriate implementation of them. While this information is unavailable for the study region, especially when considering the wind information is needed for all the SHUs in the study region. Due to this, all measures designed for wind erosion are supposed to be excluded in this study, as without the wind related information they cannot be simulated in the model with rational outcomes.

5.2.4 Applicability for Agri-Environmental Cost Estimation

Considering the agri-environmental cost estimation of measures, there are mainly two aspects needed to be noticed during the process of measure identification. The first is that whether a measure could be described with clear detailed operational procedures for cost analysis. The second is that whether all these detailed operation procedures of a certain measure could be represented with cost formulas and have data availability for cost calculation.

For each measure the clear operational procedures in a study should be consistent for eco-hydrological model simulation and for cost estimation. With operational measures, the operational procedures and other information like plant species are specific, which makes the cost analysis of the measures clear and consistent with the representation of measures in the model. On the contrary, as described above for the problem of the representation of structural measures in eco-hydrological models, it is difficult to do the cost analysis of structural measures without clear implementation procedures.

As to the data availability, the thing need to be considered is that the operational procedures of some measures might never be applied by farmers in reality in the past. These procedures are usually referred to scenarios. This leads to the situation that the related data for the cost calculation of these scenarios need to be collected through alternative methods instead of questionnaire survey with local farmers, and these alternative methods should be prepared. Each operational procedure of each identified measure need to be confirmed that the corresponding data for cost analysis of them are available with certain rational data collection methods.

5.3   Measures Identified

Given all the principles described above, twelve measures are identified for this study, as presented in Table 5.2 with their corresponding types and codes being described. These twelve measures are categorized as five different types, with one of the types is structural measures and all other types of measures are operational measures. Under each measure type, there are measures being showed along with the corresponding sub-measures. The aim of the sub-measures is to distinguish each measure further more to see the differences among them during the AES design process. For each of these sub-measures there is a code being adopted to represent the measure in this study, as showed in Table 5.2.

These measures are selected based on the criteria above. First, all these measures are easy for farmers to understand and not difficulty for farmers to implement in reality. The measures identified are the commonly recommended ones to farmers, which are not unfamiliar for farmers and with the explanation of the specification for each measure farmers should have the capability to implement them by themselves. Second, all of the identified measures are designed to be suitable for the application on cropland. Third, the measures are identified based on the characteristics and conditions of the study region, which would be helpful for improving the local situation of the NPS water pollution. Fourth, these measures are considered to have the applicability for the mitigation effect assessment with the eco-hydrological model of SWAT and have the applicability for applying the agri-environmental cost estimation. The data required for these measures to be simulated by SWAT and the data needed for them to be analyzed for cost calculation are considered to be available.

Each of these measures have its corresponding functions, which is aimed either on the individual pollutant of sediment, nitrogen and phosphorus, or on the combination of them. The explanation for each of the identified measures is described in Table 5.2.

Table 5.2 Measures identified for this study

Type	Measures	Sub-measures	Codes
structural measures	filter strip	filter strip: 5 meters	M1
		filter strip: 10 meters	M2
		filter strip: 15 meters	M3
tillage activities	no-till	no-till	M4
nutrient management	chemical fertilizer reduction	chemical fertilizer reduction by 25%	M5
		chemical fertilizer reduction by 40%	M6
	manure application	chemical fertilizer reduction by 50%, plus increasing swine manure 1000kg/ha	M7
		chemical fertilizer reduction by 50%, plus increasing sheep manure 1000kg/ha	M8
crop planting	cover crop	cover crop: soybean	M9
		cover crop: corn	M10
compounded	mixed measures	chemical fertilizer reduction by 25%, plus no-till	M11
		chemical fertilizer reduction by 40%, plus no-till	M12

Source: Own results

5.3.1 Structural Measures

The only structural measure in this study is filter strip with three sub-measures differing in terms of the width of the filter strip. Filter strip is expected to prevent the pollutants of sediments, nitrogen and phosphorus from cropland to waterbodies in the process and transport. Regarding the width of filter strip in this study, it is preferred the width to be no more than 15 meters, as in China cropland are quite small patches and scattered, and in reality in the study region many cropland patches are not wider than 15 meters. Considering this, the widths of sub-measures of filter strip are 5 meter, 10 meters and 15 meters respectively. These widths are commonly applied in other research (e.g. Arabi et al. 2006; Maringanti et al. 2011; Rodriguez et al. 2011; Parajuli et al. 2008). As to the land cover of filter strip, grass vegetation is recommended (Schmitt et al. 1999), especially the local native grass species. According to Xiao et al. (2010) and Wu et al. (2010), pennisetum is one the most promising grass species for the area of northern China, as it is native perennial which is tolerant to the local extreme climate conditions and it has sufficient strength on its stem for resisting the power of runoff.

Therefore, sub-measures of pennisetum filter strip with different widths of 5 meters, 10 meters, and 15 meters are identified in this study for the cropland that is contiguous to the river channels. It will be located at the lower edges of cropland, which is also the area contiguous to the waterbodies.

5.3.2 Operational Measures

The measure of no-till aims to reduce the disturbance of soil on cropland through the reduction of the tillage activities. The function of the measures is believed mainly to reduce soil erosion, and promote the soil health with the increased organic matter content in soil. Along with the measure of no-till there is the special no-till planter which is designed for the application of the seeding work on farmland. The measures of no-till is to skip the tillage activities of ploughing and harrowing and only do the work of seeding with no-till planters for the planting of crops. No-till practice is generally thought to induce the great growing of weed on cropland, thus the increased application of herbicides to control weeds compared to traditional farming is usually needed. In many research, no-till is explained along with the residue management activates, that is the residue of crops and other plants is appropriately covered on the surface of cropland. However, in this study the no-till measure will not consider the residue management due to the issue of data availability. Therefore, the measures of no-till here refers to the practice of adopting no-till planter for seeding without conventional ploughing and harrowing, with the increased application of herbicide for controlling more weeds at the same time.

Regarding nutrient management, there is 4Rs nutrients stewardship as explained in Table A.2 in Appendix 1 in the Electronic Supplementary Material. 4Rs nutrients stewardship refers to the careful management of the amount, source, placement, form and timing of the application of plant nutrients and soil amendments. However, it is too complicated for the SWAT model simulation and for the cost estimation. In this study, nutrient management will mainly focuses on the design for the amount and source of the nutrient for cropland. The aim is to reduce the load of pollutants of nitrogen and phosphorus from cropland. There are two kinds of measures under nutrient management in this study, which are identified related to the amount and source of fertilizer application. As showed in Table 5.2, one measure is to control the amount of chemical fertilizer application, and the other measure is to increase the application of manure along with the reduction of chemical fertilizer application.

The amount of chemical fertilizer application is criticized to be too much in China, and some research gives the recommendation for the amount of fertilizer application considering the crop yield and profitability of farmers (Ju et al. 2007; Xu et al. 2014; Zhang et al. 2018). Although manure is recommended for cropland application, too much manure application is also not environmental friendly. The combination of manure and chemical fertilizer is considered to be able to get the better result of grain yield (Information Center of the Ministry of Agriculture and Rural Affaires 2017). Based on these, for the measure of chemical fertilizer reduction two sub-measures are identified with the reduction of chemical fertilizer by 25% and 40% respectively. For the measure of manure application, sub-measures of reducing the chemical fertilizer by half and applying manure of sheep and swine respectively with the amount of 1000 kg/ha are identified in this study.

Cover crop is recommended for the seasonal protection of cropland[6] and for the soil improvement of cropland. It has the function to control soil erosion, improve soil tilth, and reduce water quality degradation through utilizing excessive soil nutrients. Leguminous species are commonly recommended for cover crops, as its special ability as nitrogen-fixing plants (Ordóñez-Fernández et al. 2018; Reckling et al. 2016). In the study region, soybean and corn are less planted by farmers in the past years due to their relatively less economic benefits, while these two kinds of crops are not unfamiliar for the local farmers. Based on this, soybean and corn are identified as the plant species respectively for the two sub-measures of cover crops in this study. Besides, in order to better achieve the function of cover crops regarding the mitigation effect of NPS pollution, fertilizer application and tillage activities are not recommended. Therefore, the practice for the two sub-measures of cover crop is to plant either soybean or corn on the conventional bare cropland in the study region during the needed seasons, without any fertilizer application and any tillage activities.

---

[6] Protection for the brown bare cropland in some special seasons.

Compounded measures in Table 5.2 are the measures which are the mixed results of the individual measures described in above. The aim is to obtain the combined effect of the individual measures. The measure of no-till is mainly for the control of sediments resulted from soil erosion. Measures of chemical fertilizer reduction with different degrees are useful for the mitigation of pollutants of nitrogen and phosphorus. The combination of these two kinds of measures are expected to control all of the three pollutants. Based on this, the two sub-measures of the compounded measure are the practice of no-till combined with chemical fertilizer reduction by 25% and 40% respectively. In practice, it would be to reduce the original chemical fertilizer by 25% or 40% for each crop planted, meanwhile using the no-till planter to directly do the seeding work without conventional plowing and harrowing work.

Mitigation Effect Assessment

6

This chapter aims to obtain the quantified mitigation effects of each measure in each heterogeneous spatial unit, through the application of eco-hydrological model simulation. First, the identification for appropriate model and the introduction for the identified model of SWAT are demonstrated. Second, the SWAT model is built for the Baishahe watershed, and is applied to simulate each of the identified measures. Third, the simulated results of measure mitigation effects are analyzed.

6.1 Eco-Hydrological Model Identification

6.1.1 Features of Eco-Hydrological Models

Watershed-scale eco-hydrological models are efficient tools for assessing the mitigation effects of BMPs regarding NPS water pollution (Kalin and Hantush 2003). Eco-hydrological models are developed to simplify and simulate the complex natural processes of all kinds of hydrological phenomena, including flooding, upland and streambank soil erosion, and contamination of water caused by agricultural chemicals. The models for NPS pollution have been developed even since the 1960s, especially in US (Shen et al. 2012).

Eco-hydrological models involve knowledge of an integrative science which studies the relationships between hydrological, biogeochemical and ecological

Supplementary Information The online version contains supplementary material available at https://doi.org/10.1007/978-3-658-41340-8_6.

processes in soil, rivers and lakes, and at the catchment scale. It proposes a "dual regulation" of a system by simultaneously studying hydrological and ecological processes (Krysanova and Arnold 2008). Hydrological factors could determine the dynamics of natural and human-driven terrestrial ecosystems, and ecological factors influence water dynamics and water quality. River basin represents a suitable scale for integrated eco-hydrological studies and modelling, as it has a hierarchical structure and natural boundaries, which could be considered as inherent integrators of the effects of many climatic and non-climatic factors, like human-driven alterations (Krysanova and Arnold 2008).

Nowadays there are many eco-hydrological models available. As said by Borah and Bera (2004, p. 789) "it is difficult to choose the most suitable model for a particular watershed to address a particular problem and find solutions", as "differences are often subtle between many of these models" (Tuo et al. 2015, p. 2852). According to Borah and Bera (2004), the most suitable model for an application could mainly depend on the focused problem, watershed size, desired spatial and temporal scales, expected accuracy, user's skills, computer resources, etc. Therefore, it is important to first understand the different features of different eco-hydrological models.

(I) Employed algorithms

A watershed model may be designed based on a conceptual, empirical or physical framework (Aksoy and Kavvas 2005; Daniel et al. 2011). A brief summarized characteristics of these three different kinds of models are shown in Table 6.1, along with their corresponding features and example models.

Table 6.1 Differences of algorithms among three kinds of watershed models

	Conceptual models	Empirical models	Physically based models
Algorithm	relatively qualitative	statistical relationships by regression analysis	laws of conservation of mass, momentum, and energy
Features	clear causal relationship	less input data, simpler calculation	complete system, long-term, clearer spatial characteristics
Examples	HBV, LASCAM	USLE, AGNPS	SWAT, HSPF, MIKE SHE

Source: Content adapted from Aghakouchak and Habib (2010), Kalin and Hantush (2003), Daniel et al. (2011)

In conceptual models, some simple mathematical equations, rather than governing partial differential equations, are used to describe the main hydrologic processes, such as evapotranspiration, surface storage, percolation, snowmelt, base flow, and runoff (Aghakouchak and Habib 2010). They can be viewed as qualitative and descriptive statements of hypotheses concerning the nature of the potential causal relationship between activities and their impacts (Reiter et al. 2009). Most conceptual models consider the catchment as a single homogeneous unit (Solomatine and Wagener 2011). Examples of conceptual model include HBV and LASCAM (Aksoy and Kavvas 2005; Aghakouchak and Habib 2010; Solomatine and Wagener 2011).

Empirical models are based on statistical relationships obtained through regression analysis of observed data (Kalin and Hantush 2003). These data are monitored water quality and quantity in runoff from typical experimental plots, and are used to build empirical relationships between hydrological parameters. As it seldom takes the complex pollution process and mechanisms into consideration, it is known as "black box" research methods (Shen et al. 2012). The advantages for this kind of modes are relatively lower demand of input data and simpler calculation process, while disadvantages refer to the limitation to conditions for which they are developed. They cannot describe the contaminant migration process well, thus are less reliable under the conditions outside the limit of the original environment and generally are not suitable for predictions under different conditions (Shen et al. 2012; Kalin and Hantush 2003). Examples of empirical model include USLE, MUSLE, RUSLE, SEDD, AGNPS (Aksoy and Kavvas 2005).

Physically based models, in contrast to empirical models, are based on laws of conservation of mass, momentum, and energy, and use much more detailed and rigorous representations of these physical processes (Solomatine and Wagener 2011). It is a relatively complete model system associated with all kinds of other specific models, including hydrological, erosion and pollutant migration ones (Shen et al. 2014b). Known as "white-box" research methods, physically based models can calculate the occurrence of NPS pollution quantitatively in a continuous long-term process and have clearer spatial characteristics distributions (Shen et al. 2012). As a result of improvement in computer power, this kind of model became practically applicable in 1980s (Solomatine and Wagener 2011). However, the disadvantages of physically based models are requirements for a large body of input data and difficulties on calibration and validation with limited available information, due to the large number of parameters. Some widely accepted physically based models include SWAT, HSPF, MIKE SHE (Daniel et al. 2011).

(II) Temporal variability

From the perspective of temporal scale, watershed models can be subdivided into short-duration storm event-based and long-term continuous ones. Examples of normally mentioned event-based watershed models include DWSM, AGNPS, ANSWERS; and continuous watershed models include HSPF, SWAT, AnnAGNPS, APEX.

Event-based models are designed to simulate individual storm or rainfall-runoff event. It is useful for analyzing severe storm events and evaluating watershed management practices, especially structural BMPs (Borah and Bera 2003). It can be of particular interest resulted from the reason that generally it is the intense storms which cause flooding and carry most of the yearly loads of sediment and pollutants (Borah and Bera 2004). Continuous model is used for the simulation of many consecutive rainfall-runoff events occurring during a season or longer time period. It can explicitly account for all runoff components while considering soil moisture redistribution between storm events (Daniel et al. 2011). Continuous models are commonly adopted for analyzing effects of long-term hydrological changes and BMPs, especially agricultural BMPs. Considering the land use suitability, among the long—term continuous simulation models, SWAT is promising in predominantly agricultural watersheds, HSPF is promising in mixed agricultural and urban watersheds (Borah and Bera 2004).

(III) Spatial variability

On a spatial basis, watershed models can be categorized as lumped, semi-distributed, and distributed models. Examples for lumped models include EPIC and CLEAMS, and most of other commonly applied models are semi-distributed or distributed ones, such as SWAT, AnnAGNPS, APEX. Lumped modeling approach considers the whole watershed as a single unit with no spatial variability, using single values for all kinds of input parameters and resulting in single outputs. On the contrary, semi-distributed and distributed models take account for the spatial variability regarding hydrologic processes, data input, boundary conditions, and watershed characteristics. By doing this, spatially distributed parameters are adopted for input and spatially distributed outputs are provided. The difference between semi-distributed and distributed models is mainly the different degrees regarding the spatial heterogeneity. In the literature, terms for semi-distributed and distributed models are not always distinguished, and distributed models can be a general representation for semi-distributed ones.

For the application in practices, distributed models are generally favored over lumped models regarding the detailed total maximum daily load (TMDL) development and the implementations of BMPs (Kalin & Hantush, 2003). Among the distributed models, the physically-based ones are generally preferred over empirical ones, since model parameters have physical meaning and can be measured in the field or deduced from published data in the literature. However, most physically based and distributed models require enormous amount of input data.

(IV) Major components
Eco-hydrological modeling involves five natural compartments, including atmosphere, soil, surface water, groundwater and vegetation (Tuo et al. 2015). Different models approach the simplification of this complex system in different ways, with their different constituted components and the techniques behind them. The normally referred components include hydrology, sedimentation, nutrients, pesticides, crop growth, agricultural management, reservoir and channel routing, etc.

Generally, the analysis of the components of a model can be recognized as the analysis of the capability of the model. There are three major components acted as the main processes in the model to determine the characteristics of NPS pollution, which are the ones for hydrology, sediment, and chemical (Hu et al. 2002). When identifying promising models for applications, these three major components are emphasized for the analysis (Borah and Bera 2003, 2004). Models that are equipped with all these three major components include AGNPS, AnnAGNPS, DWSM, HSPF, MIKE SHE, and SWAT. These three components are generally reflected in the NPS pollution models by their respective modules, with the relationship between them being illustrated in Figure 6.1. Hydrological process is the major driving force for sediment generation and transport as well as for chemical pollutants transportation, meanwhile chemical pollutants can also be transported in the means of attaching to sediment.

Under each module in Figure 6.1, there are subdivided components represented the sub-processes in the model. Among the three major components, hydrology constitutes the most important one (Kalin and Hantush 2003), as water power is the driving force for the movement of sediments and chemical pollutants. According to Tuo et al. (2015), SWAT has the most comprehensive hydrological processes considered among all the five selected physically based models (i.e. SWAT, AnnAGNPS, HSPF, SWIM, GWLF).

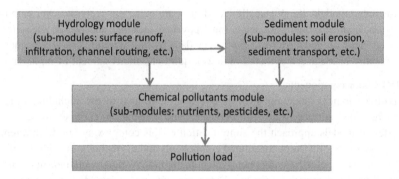

Figure 6.1   Structure of typical NPS models (Source: Modified from Hu et al. (2002))

6.1.2   Suitability

First, the aim to apply an eco-hydrological model in this study is to simulate the study region as accurate as possible and get quantified results of the mitigation effects of BMPs. Therefore, instead of the conceptual and empirical models, physically based models which consider the complex pollution process and mechanisms, have quantitative analysis, and are able to simulate various environmental conditions are required. Second, this study emphasizes the spatial heterogeneity for the aim of cost-effective analysis. As demonstrated by Veith et al. (2004), cost-effectiveness of NPS pollution reduction programs in an agricultural watershed depends on the selection and placement of control measures within the watershed. Due to this, semi-distributed or distributed models which account for the spatial variability are needed. Third, the yearly average results of the mitigation effects of BMPs within long-term period are demanded in this study from the model. This induces that the long-term continuous models are preferred compared to short-term event-base models. Fourth, the study region to be simulated by the eco-hydrological model is agricultural area, with the targeted pollutants being sediments, nitrogen and phosphorus. Therefore, the potential model must be suitable for agricultural area, and be equipped with all the three major components as showed in Figure 6.1. Last but not least, the identified model in this study is expected to have high capability for the simulation of different kinds of BMPs.

Based on these, the criteria for identifying the appropriate eco-hydrological model in this study are summarized:

I.   physically-based;
II. long-term continuous in terms of temporal variability;
III. distributed or semi-distributed in terms of spatial variability;
IV. have three major components: hydrology, sediment, chemical;
V.   suitable for agricultural watershed;
VI. high capability for the simulation of all kinds of BMPs.

With these criteria, SWAT model is identified for this study. According to Giri et al. (2014), among many models (include SWAT, HSPF, AnnAGNPS, ANSERS-2000 and WEPP), SWAT is preferred as its built-in equations describe various agricultural components (e.g. tillage operation, fertilizer application, crop rotation, vegetative filter strips) in more detail than others for BMPs simulation. As reviewed by Kalin and Hantush (2003) regarding the simulation capability of BMPs, among many models (including AnnAGNPS, ANSWER2–2000, HSPF, MIKE-SHE, SWAT and others) SWAT and AnnAGNPS are concluded to be suitable for BMPs simulations aimed at sediment and nutrients in agricultural areas. Xie et al. (2015) demonstrates that SWAT has the most comprehensive BMP simulation capability under the comparison of many other models (AGNPS, AnnAGNPS, HSPF). In addition, SWAT model has the component of crop growth, which could be very helpful to obtain the spatially heterogeneous data of crop yield from the model for cost estimation. The component of crop growth is rare for many other models, such as AnnAGNPS.

6.2 SWAT Overview

6.2.1 Development History

SWAT, the Soil and Water Assessment Tool, is a physically based, long-term continuous and semi-distributed eco-hydrological model that operates on a daily time step in the river basin scale. It is initially designed to simulate the effects of alternative practices on water, sediment, and agricultural chemical yields for ungauged rural basins across the USA. The first version of SWAT is created in the early 1990 s, by merging other previously developed models (Krysanova and White 2015), as showed in Figure 6.2. It is a period of over 30 years' modelling experience for USDA Agricultural Research Service to lead to SWAT (Arnold et al. 2012).

In the mid-1970s, CREAMS was developed, which was process-based, NPS model for the impacts of land management on water, sediments and nutrients in field scale. In 1980s, several new models were developed, including EPIC for the impacts of erosion on crop productivity and GLEAMS for pesticide and nutrient loads to groundwater. Afterwards, SWRRB emerged for dividing watersheds into multiple sub-basins, based on the modification of components of CREAMS, EPIC, GLEAMS, and the addition of several new components. In the early 1990s, SWAT developed by incorporating SWRRB, ROTO, in-stream kinetic routines from QUAL2E model, and GRASS (Geographic Resources Analysis Support System) GIS interface (Arnold et al. 2012; Krysanova and Arnold 2008; Gassman et al. 2007), as showed in Figure 6.2.

Since then, SWAT has undergone continued review, modification, and expansion of capabilities with developing versions over the past decades. In addition, other models emerged based on SWAT for specific applications, including SWIM (aiming at climate and land-use change impact assessment in urban areas), SWAT-G (focused on flow prediction for low mountain ranges in Germany), ESWAT (for hourly time step), and SWAT-MODFLOW (coupled watershed and groundwater model).

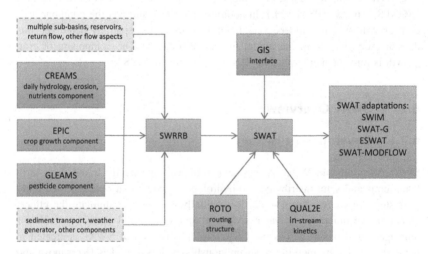

Figure 6.2 Schematic of SWAT development history (Source: Modified from Arnold et al. (2012), Krysanova and Arnold (2008), and Gassman et al. (2007))

6.2.2 Spatial and Temporal Dimensions

The way a watershed is discretized determines the basic computational units in which certain types of BMPs are simulated (Xie et al. 2015). As a semi-distributed model, SWAT discretizes a watershed first into subunits including sub-basins, main channel segments, impoundments on the main channel network, and the inclusion of measured data from point sources in some sub-basins (Arabi et al. 2008). Sub-basins can be further subdivided into HRUs that are portions of sub-basins consisting of homogeneous land use, management, and soil characteristics. The HRUs are represented as a percentage of the sub-watershed area and are not spatially identified within a SWAT simulation (Jha et al. 2004). HRUs are not always necessary for the delineation of the watershed, and a watershed can be only subdivided into sub-basins that are characterized by dominant land use, soil type, and management.

As a long-term continuous model, SWAT operates on a daily time step. Correspondingly, the daily weather data are needed for SWAT model set up. For SWAT model simulation, a warm up period of at least two to three years is demanded. The warm up period is important for SWAT model to generate and approach reasonable initial values for model variables, such as the soil water content and surface residue (Shukla 2011; Zhang et al. 2007). The results of SWAT model simulation can be received based on daily, monthly, or annual time scales depending on the demand (Tuo et al. 2016).

6.2.3 Components

The major components of SWAT include hydrology, weather, sedimentation, soil temperature and properties, nutrients, pesticides, bacteria and pathogens, plant growth, and land management (Arnold et al. 2012; Gassman et al. 2007; Borah and Bera 2003).

Hydrology is the most basic and important component for SWAT. Water balance is the driving principle behind all the processes in SWAT, which impacts the process of pant growth and the movement of sediments, nutrients, pesticides, and pathogens. There are many sub-processes which are integrated into the hydrological component in SWAT, including canopy storage, surface runoff, infiltration, evapotranspiration, lateral flow, tile drainage, redistribution of water within the soil profile, consumptive use through pumping, reservoirs, ponds, and tributary channels (Yang et al. 2013a; Arnold et al. 2012). The simulation of hydrology of a watershed with SWAT is separated into two phases, i.e. the upland phase and the in-stream phase (Krysanova and Srinivasan 2015). The upland phase mainly controls the amount of flow water, sediment, nutrient, and pesticide loadings in each sub-basin to the main channel (Figure 6.3). The in-stream phase mainly controls the movement and transport of these upland loadings from each sub-basin through the channel network to the outlet of the watershed.

The hydrologic cycle within the upland phase in the model is driven by climate conditions, which provide moisture and energy inputs (Yang et al. 2013a). Climate conditions in SWAT refer to five aspects of daily precipitation, maximum and minimum air temperature, solar radiation, wind speed, and relative humidity. Daily climate data regarding these five aspects is needed for SWAT model set up. These climate input will determine the relative importance of the different components of the hydrologic cycle, as showed in Figure 6.3. The daily water budget in each HRU is computed based on daily precipitation, runoff, evapotranspiration, percolation, and return flow from the subsurface and ground water flow.

Soil properties are important for SWAT model, as they can control the infiltration and soil water movement and play a key role in surface runoff, groundwater recharge, evapotranspiration, soil erosion, and the transport of chemicals within the hydrologic cycle (Yang et al. 2013a). Soil temperature refers to the different levels of temperature in the different depths of the soil underground, which will impact the soil formation, plant growth, and the hydrologic cycle. In SWAT there is a specific formula for the calculation of soil temperature based on the different air temperature, soil depths and other factors. Up to ten soil layers can be divided in SWAT, users can decide the number of soil layers depending on the available data of the research region. For each layer both the physical and chemical features for the soil will be described with all kinds of soil parameters. Sediment yield resulted from water erosion is simulated in SWAT with the Modified Universal Soil Loss Equation (Arabi et al. 2008).

For nutrients (mainly nitrogen and phosphorus) and pesticides, in SWAT model there are many compound forms for each of them[1] regarding the simulation of movement and transformation. The loss of nitrogen and phosphorus from the soil system in SWAT can occur by crop uptake as well as by surface runoff through the solution in water and the attachment on eroded sediments (Gassman et al. 2007). Simulated loss of nitrogen can also occur in percolation below the root zone, in lateral subsurface flow including tile drains, and by volatilization to the atmosphere. Regarding the pesticide transport and loss, the processes include the degradation and loss by volatilization, leaching, and attachment on eroded

---

[1] Like for nitrogen, the existed forms of NO_3, NH_4, NO_2 are considered in SWAT.

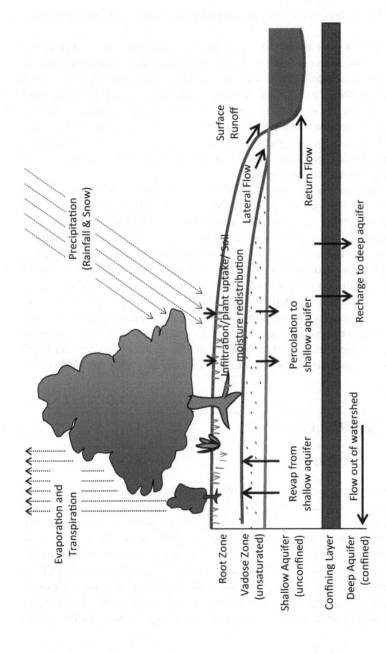

Figure 6.3 Schematic of SWAT land phase hydrologic cycle (Source: Modified from Yang et al. (2013a))

sediment, as well as in the solution phase of surface runoff and lateral subsurface flow.

Plant growth sub-model in SWAT can be adopted to assess the crop yields and biomass output, as well as water and nutrients uptake and transformation from the root zones inside each HRU with its corresponding land cover. The plant growth sub-model in SWAT is a simplified version of the plant module from the model of EPIC, which can simulate all types of land covers and differentiate between annual plants and perennial plants (Arnold et al. 2012; Krysanova and Arnold 2008). A wide range of agricultural land use arrangements, such as cropping rotations, grassland or pasture systems, and trees, can be estimated for obtaining the crop yields and biomass outputs.

Regarding land management, SWAT allows the user to define management practices taking place in each HRU (Arnold et al. 2012). It can simulate all kinds of agricultural activities, such as cropping system of planting, harvesting, tillage passes, nutrient applications, and pesticide applications, with the means of either specific dates input or a heat unit scheduling approach. Nutrient application can be simulated in forms of chemical fertilizer input and organic fertilizer input; meanwhile, an alternative of auto fertilizer routine can be used to simulate nitrogen applications. Similarly, irrigation application can be applied either with specific dates and amounts or with the auto irrigation option. A number of agricultural land management practices can be simulated in SWAT, such as many of the BMPs. Some standardized methods for simulating these practices with SWAT model are presented by researchers, such as Arabi et al. (2008) and Waidler et al. (2011).

6.3   SWAT Setup

6.3.1   Data Preparation

There are extensive data need for SWAT model set up. As the model involves five natural compartments, i.e. atmosphere, soil, surface water, groundwater and vegetation (Tuo et al. 2015), the data requirement mainly include topography, soil, climate, land use, and land management, and others. These data can be catalogued into two types. One is spatial data, in the format of raster, with the necessary spatial data for SWAT including DEM (Digital Elevation Model) for topography, soil map, and land use map. The other kind of data is shown as attribute table, with the required ones including soil attribute table, land use type attribute table, climate database, and data for land management. Other optional data for SWAT model setup include reservoir data, point source pollution data, river network map, etc. For the study area of the Baishahe watershed in this thesis, there is no reservoirs inside and no firms as the point pollution sources. River network maps are usually offered by government for big and important basins, like Yellow River, which is used to adjust the drainage network resulted from DEM. It is not available for this study, due to the small area of the Baishahe watershed.

Based on this, the overview of the modelling data of SWAT in this study is summarized in Table 6.2, with the explanation for data sources and pre-processing in the following.

Table 6.2 Overview of the data for SWAT setup in this study

Type	Name	Source (version)	Resolution	Original year	Released year
Spatial data	DEM	Geospatial Data Cloud (ASTER GDEMV2)	30 m × 30 m	2000–2010	2015
	Soil	Harmonized World Soil Database (Version 1.21)	1 km × 1 km	1995	2012
	Land use	Data Center for Resources and Environmental Sciences, Chinese Academy of Sciences	30 m × 30 m	2015	none
Attribute data	Climate	China Meteorological Assimilation Driving Datasets for the SWAT model (Version 1.1)	1/4° × 1/4° (daily, 2008~2016)	2016	2018

(continued)

Table 6.2 (continued)

Type	Name	Source (version)	Resolution	Original year	Released year
	Land management	personal communications with the local government officers and farmers in 2018			
	Soil type	same as the spatial data of soil above			
	Land use type	same as the spatial data of land use above			

Source: Own results

The data source of DEM is Geospatial Data Cloud (Geospatial Data Cloud 2015). The spatial resolution of the data is 30 meters, as showed in Table 6.2. The data was collected under the technique of ASTER[2], and was first released by METI[3] of Japan and USA NASA[4] in 2009 with the first version of ASTER GDEMv1. The DEM for this study is the modified and improved second version of the data which was released in 2015, i.e. ASTER GDEMv2. The dataset of ASTER GDEMv2 was generated from ASTER instrument acquired between 2000 and 2010, with the format of GeoTIFF (.tif). After the pre-processing of projection transformation[5] and clipping boundary with ArcGIS, the DEM adopted for this study is gained, as showed in Figure A.4 in Appendix 2 in the Electronic Supplementary Material. The land elevation of the Baishahe watershed ranges from 574 meters to 1577 meters (Figure A.4).

The soil data is obtained from the global soil database of Harmonized World Soil Database (HWSD) (FAO/IIASA/ISRIC/ISSCAS/JRC 2012). It is a global soil database established jointly by IIASA and FAO in partnership with ISRIC-world Soil Information, the European Soil Bureau Network and the Institute of Soil Science, Chinese Academy of Sciences (van Velthuizen 2017; Amundson et al. 2015). The database is composed of a 30 arc-second raster image file and a linked attribute database, covering the globe's land territory. For Chinese territory, the soil data is provided by the Institute of Soil Science in Chinese Academy of Sciences based on year 1995 with a 1:1,000,000 scale. The resolution of the soil date is approximately 1×1 km, as notified in Table 6.2. Regarding soil data in SWAT, there are two kinds required, which are the spatial map of soil and

---

[2] The Advanced Spaceborne Thermal Emission and Reflection Radiometer.

[3] The Ministry of Economy, Trade, and Industry.

[4] The United States National Aeronautics and Space Administration.

[5] Converting with the Geographic Coordinate System Transformations in ArcGIS from "GCS_WGS_1984" to "GCS_New_Beijing".

the attribute data of the corresponding soil types. The soil map of the Baishahe watershed is obtained after the data pre-processing of projection transformation[6] and clipping boundary with ArcGIS, as illustrated in Figure A.1 in Appendix 2 in the Electronic Supplementary Material. There are two types of soil in the Baishahe watershed, each of which needs attribute data input for SWAT modelling. The obtaining of soil attribute data need complex analysis and calculation, which are demonstrated in detail in Appendix 2 in the Electronic Supplementary Material, with the results being showed in Table A.8 in Appendix 2 in the Electronic Supplementary Material.

The land use data is gathered from the Data Center for Resources and Environmental Sciences, Chinese Academy of Sciences (Resources and Environmental Science Data Center 2015b). The data is the latest version collected in 2015 and based on the version of 2010, with a resolution of 30×30 meters, as showed in Table 6.2. Regarding the spatial distribution map of land use, the pre-processing includes projection transformation[7], clipping boundary with ArcGIS, and land use reclassification. As the land use data for this study are based on Chinese system (as shown in Table 6.2), while the SWAT model is designed based on the situation of US, the reclassification of land use with the original data is performed in order to be consistent with the classification of land use for SWAT model. The detailed process for reclassification is demonstrated in Appendix 2 in the Electronic Supplementary Material, with the final spatial distribution of land use map in the Baishahe watershed being showed in Figure A.3-B in Appendix 2 in the Electronic Supplementary Material. After the reclassification, there are three kinds of land use in the study region, with the attribute data for each of them being consistent with the ones along with SWAT model.

Regarding climate data, daily data are required for SWAT modelling with five climate aspects, as described in Table 6.3. For this study, the China Meteorological Assimilation Driving Datasets for the SWAT model (CMADS) is gathered and applied as the climate input (Meng 2016). The collected data include the daily data for all the five climate indices from 2008 to 2016 in the format that SWAT model required. CMADS is a new dataset developed by Meng et al. (2017), which is based on the China Meteorological Administration Land Data Assimilation System assimilation technology with multi-source data such as satellite observation, land surface observation and numerical products. It is built with the

---

[6] Converting with the Geographic Coordinate System Transformations in ArcGIS from "GCS_WGS_1984" to "GCS_New_Beijing".

[7] Converting with the Geographic Coordinate System Transformations in ArcGIS from the original form to the form of "GCS_New_Beijing".

methods of data loop nesting, resampling and bilinear interpolation. CMADS covers the East Asia area (between 0 N and 65 N, 60E and 160E), with a spatial resolution of $1/4° \times 1/4°$, consisting of 300×195 grid points for totally 58500 stations. For SWAT model input, each HRU in the watershed searches its nearest weather station to get the climate data, therefore all the related stations need to be extracted according to the their locations compared to the study region. For the Baishahe watershed, due to the small area of it, there will be only one station needed which is the nearest for all the possible HRUs, as the map of study region and weather stations being showed in Figure A.5 in Appendix 2 in the Electronic Supplementary Material.

Table 6.3 Climate data requirement for SWAT

Climate indices	Unit
Daily precipitation	mm
Daily maximum and minimum temperature	°C
Daily solar radiation	MJ/m^2
Daily relative humidity	in fraction format
Daily wind speed	m/s

Source: Content adapted from Yang et al. (2013a)

The information of the initial land management situation in the watershed is needed for the design of the.mgt file in SWAT modelling. These are important for reasonably simulating runoff, sediment and water quality processes (Yang et al. 2013a). Land management information regarding cropland includes planting date, harvest date, irrigation events, nutrient application dates and rates, pesticide application dates and rates, tillage operations and timing, and others. These information is mainly gathered from the local government officers and the representative farmers, such as the leaders of villages (local farmers, personal communication, July, 2018; J.J. Jin, C. Guo, Water Conservancy Bureau, Xia county, personal communication, January, 2018). The detailed information for cropland activities along with the time during a year is summarized in Table A.10 in Appendix 2 in the Electronic Supplementary Material.

6.3.2 Model Configuration

To setup the model project for the Baishahe watershed, the SWAT model version 2012 with the ArcSWAT interface is applied.

(I) Watershed delineation
The first step for SWAT model setup is watershed delineation, which involves delineating stream networks and sub-watersheds. This work is based on the DEM data for the study region. To delineate the boundary of the study watershed, a final outlet[8] for the whole watershed needs to be defined. In this study, it is defined as the location of the monitoring station of water quality which is set by the Water Conservancy Bureau in Xia county, with the east longitude and northern latitude being 111°16'46.9" and 35°4'57.2" respectively. The number of sub-basins and the density of the stream networks are controlled by the parameter value of Critical Source Area[9], which is set as 25 ha in this study. Based on the information described before, there are no point source and reservoirs to be added in this watershed.

Generally, with the information of the locations of the final watershed outlet, point sources of pollution, reservoirs, as well as the value of Critical Source Area, a watershed can be delineated with the result of sub-watersheds and stream tributaries based on the DEM data through ArcSWAT. Regarding the Baishahe watershed, combining with the land-use map, the area of pasture is very small and scattered, which induces that the delineated sub-watersheds might have no representatives for the land use of pasture if followed the automatic delineation process[10]. To make the delineated sub-watersheds to be more realistic regarding the land use, the function for adding and deleting outlets of sub-watersheds is applied with the delineation tool. For each sub-watershed there is an outlet for that area, which is the outlet of the stream tributary inside that sub-watershed. As showed in Figure 6.4, the manually added outlets are noticed as the green points. In this process, some adjacent sub-watersheds with the same land use of forest are merged by deleting relevant outlets, as the study focuses on the cropland.

As a result, in Figure 6.4 a watershed with 55 km^2 is illustrated, with totally 83 sub-watersheds being defined. The final outlet of the whole watershed is located

---

[8] An outlet refers to a (relatively) lowest point in the downstream where water in the upstream will flow into there.

[9] It refers to the minimum area of water concentration which is assumed to be set as stream flow in the model.

[10] As in this study, each sub-watershed is designed as a homogeneous spatial unit in terms of land use, soil, climate and others.

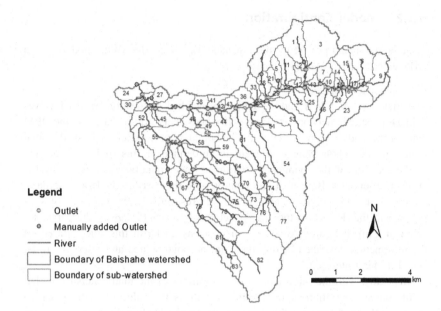

Figure 6.4   Delineation result of Baishahe watershed (Source: Own results from ArcSWAT)

in the westernmost sub-watershed with the code of 24. The shapes and boundaries of the delineated river and sub-watersheds are illustrated. Besides, the characteristics and parameter values for each stream tributary and sub-watershed can be checked in the attached report of the model delineation. Regarding characteristics of stream tributaries, it includes the length, slope, and bank full width and depth. For sub-watershed parameters, it includes the area, average slope, mean elevation and others.

(II) Soil, land use input and slope discretization
After watershed delineation, the spatial distribution maps of soil and land use and their attached attribute data need to be input into the model respectively. These two maps (Figure A.1-B and Figure A.3-B) will be overlapped with the watershed delineation map above (Figure 6.4). Afterwards, for each sub-watershed the distribution of soil and land-use can be checked and calculated in terms of area through the model. These soil and land use are simulated in the model based on

their attached attribute data (i.e. parameters). As described in the data preparation part, there are totally threes kinds of land use and two kinds of soil in the Baishahe watershed.

The slope values were extracted from the DEM data. There are two options in SWAT for spatial slope discretization with the functions of "single slope" and "multiple slopes". Single slope refers to an average slope value being resulted for each sub-watershed from the model. Multiple slopes means that each sub-watershed could be differentiated with maximum five slope classes. In this study, the multiple slope discretization operation is performed with five slope classes, as the region is mountainous and the sub-watersheds differs greatly in terms of slope. The resulted slope map of the Baishahe watershed is shown in Figure A.6 in Appendix 2 in the Electronic Supplementary Material, with the minimum, maximum, and mean slope in the region are 0, 185.52, and 35.32 respectively (in the form of percentage).

(III) HRU definition

SWAT can further divide each sub-watershed into one or more HRUs. HRUs are essentially smaller land units with each HRU having homogeneous soil, land use and slope (Moriasi et al. 2019). However, in a sub-watershed with multiple HRUs, the exact spatial location of each HRU cannot be defined by SWAT; each of HRUs is represented with its area percentage in the sub-watershed. The advantage for subdividing a sub-watershed into multiple HRUs is that the model can reflect differences in runoff, erosion, nutrient loading and other hydrologic processes for different land covers and soils in more detail. However, a key weakness for SWAT is also showed here as the non-spatial aspects of HRUs, which means that the model is difficult to determine the spatial locations and describe the interactions between different HRUs (Ning et al. 2015).

Based on these, after the overlapping of land-use, soil maps and slope discretization, the definition for HRUs in the modelling is performed as that setting only one HRU in each sub-watershed with the option of "Dominant Land Use, Soil, Slope". It means that one sub-watershed will represent one HRU, with the defined homogeneous features of soil, land use, and slope being the one that is dominated in terms of area in the sub-watershed respectively. As a result, after HRU definition, the model gets a total number of 83 HRUs, with 50 for cropland, 6 for pasture and 27 for forest, as showed in Figure 6.5. The statistics regarding to the different sub-watershed numbers, areas and proportions in terms of area for different land uses are described in Table 6.4. The distribution in a visible way for this is shown in Figure 6.5.

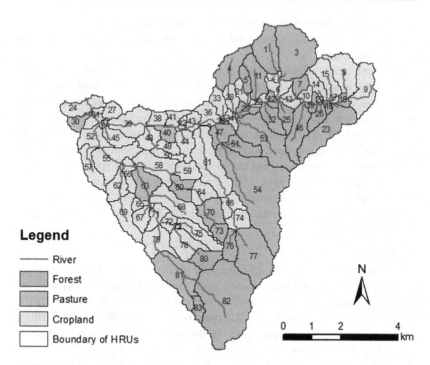

Figure 6.5 Reclassification of land-use map after HRU definition (Source: Own results from ArcSWAT)

Table 6.4 Baishahe watershed distribution situation after HRU definition

Land use	Number of HRUs	Area (km^2)	Proportion of area (%)
forest	27	25.2	45.85
pasture	6	4.3	7.79
cropland	50	25.5	46.36
in total (whole watershed)	83	55	100

Source: Own results

(IV) Land management definition

According to the information gathered for the local cropping situation in the study watershed (described in above), a distribution map for cropland in the watershed with two types of cropping system is shown in Figure 6.6. Sub-watersheds with blue color demonstrates that both winter wheat and corn are planted on the cropland in different time in a year, which is mainly performed at the local region on cropland patches closing to the river and there is irrigation activities along with it. Sub-watersheds with yellow color in Figure 6.6 represent the cropland patches which only have winter wheat being planted during a year, and these cropland are usually far away from the river without any irrigation activities being performed.

For each sub-watershed covered with the land use of cropland, the .mgt file in SWAT model which refers to the definition of detailed operations of cropping need to be edited according to the local situation. Regarding each type of cropping system in Figure 6.6, the information of land management activates ordered according to time is organized in Table A.10 in Appendix 2 in the Electronic Supplementary Material, along with their corresponding sub-watershed codes. The land management definition for the .mgt file in SWAT model is performed based on the information in Table A.10 in Appendix 2 in the Electronic Supplementary Material.

6.3.3 Calibration and Validation

After the configuration of SWAT model, the study region could be simulated by the model. However, the performance of the model is unknown, and could be checked and improved through the procedures of model calibration and validation. Model calibration is an effort to better parameterize a model for getting more precise simulated results according to the reality of the study region. The procedure is performed as to adjust the values of relevant parameters with the aim of optimizing the agreement between observed data in reality and model simulation results (Yang et al. 2013a). Model validation aims at checking if the parameterization result from calibration can work well. It compares the simulated results of model after calibration with observed data in reality in the time scope outside of the calibration period. A good validation result shows that the

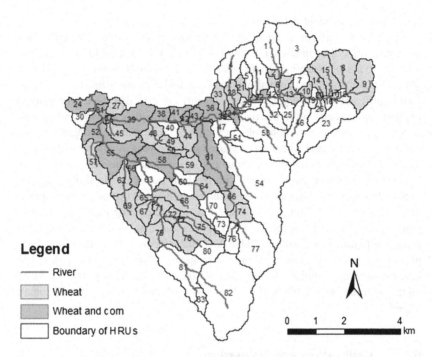

Figure 6.6 Distribution of two kinds of cropland in the watershed (Source: Own results from ArcSWAT)

calibrated model is a good simulator and does not over-fit[11] in the calibration period.

For processing the model calibration and validation, the first step needed is to obtain the most sensitive parameters for a model regarding the simulated study region. Regarding sensitive parameters, a small change of their values could have a result of a big change of the model output. During the calibration process, only the most sensitive parameters are adjusted within an acceptable value range, while other parameters remain at their default values. There are generally two ways to decide the sensitive parameters for a study, with one based on expert experience and the other according to sensitive analysis. The ways to do the sensitivity

---

[11] A model is over-fitting if it makes good predictions on a test set but bad predictions on new data.

analysis for SWAT model include One-factor-At-a-Time analysis and Global Sensitivity analysis. One-factor-At-a-Time approach is performed by changing only one parameter's value at one time manually. Global Sensitivity is conducted by the change of the values of a various of parameters at one time with the auto calibration tool of SWAT-CUP[12].

The aim of SWAT modelling in this study is to simulate the quantitative results regarding the pollutants of sediments, nitrogen and phosphorus. These three indices are the interest of calibration and validation objectives. According to Arnold et al. (2012), the recommended sequence for the calibration of these indices is first streamflow, then sediments (or together with streamflow), and finally nutrients (or together with streamflow and sediments). The main reason is that streamflow as the water power is the driving force for all other pollutants, and sediments could be the carrier of nutrients (as illustrated in Figure 6.1). Besides, the quantitative results of crop yield from SWAT are also needed in this study regarding the task of agri-economic cost calculation. When the calibration of crop yield is also demanded, it is better to conduct it before the calibration of streamflow. As some of the plant growth parameters influence the factors related for streamflow, such as evaporation, infiltration and others (Sun and Ren 2013).

The work of calibration and validation can be hindered by the limitation of observed data from reality. In this study, observed data for nitrogen and phosphorus in the study watershed are not available. Calibration and validation are processed first for crop yield, and then for streamflow and sediment.

(I) Crop yield
Regarding the observed data for crop yield calibration, the data are gathered from Yuncheng Statistical Yearbook for the two main crops in the local area (corn and winter wheat), as attached in Table A.12 in Appendix 2 in the Electronic Supplementary Material. The area of Yuncheng is about 14,000 km^2, compared with only 55 km^2 of the study watershed this statistical data would be only helpful for giving the reference of the relatively value levels of the main crop yield, instead of comparing the exact data for doing the calibration. As the administrative boundary is the way for government to do the statistic, while the study region is a watershed where it is hard to get the statistical crop yield data only

---

[12] SWAT-CUP (Calibration and Uncertainty Procedures) is a standalone program specially developed for the calibration of SWAT model, which contains five different optimization procedures for calibration and includes functionalities for validation, sensitivity analysis and uncertainty analysis (Abbaspour et al., 2015).

for that area. Besides, different planting patterns can impact crop yield a lot. The corn is planted in the study region mainly as the cover crop, which is planted during the period when the land had no wheat growing and usually the time is not enough for it to have good yield. Therefore, the statistical data for the yield of corn from the local Yearbook have less reference for the calibration of corn yield. Given all of these, manual calibration, which could make the crop yield simulated by the model to be generally consistent with the yield levels in reality in the study region, would be suitable. Meanwhile, the calibration will only be performed for winter wheat.

The sensitive parameters for crop yield mainly refer to the plant growth module in SWAT. The module is a simplification of the Environmental Policy Integrated Calculator crop model, which is used to simulate leaf area development, biomass accumulation, and crop yield for different plant species (Sun and Ren 2013). The potential sensitive parameters for the crop yield of winter wheat are obtained based on previous research (Sun and Ren 2013), which focuses on the situation of North China where the Baishahe watershed is located. The recommended parameters are tested in this study, with the final adjusted parameters and their corresponding calibrated values being displayed below in Table A.13 in Appendix 2 in the Electronic Supplementary Material. All of these parameters are sensitive to the output of crop yield in the model, and all of them are located in the CROP.DAT file in SWAT model.

(II) Streamflow and sediment

The observed data for the calibration and validation of streamflow and sediments are from the Water Conservancy Bureau in Xia county, with a monthly time step from 2009 to 2012. These data are sorted out and described in Table A.11 in Appendix 2 in the Electronic Supplementary Material. Observed data in 2009 and 2010 are adopted for model calibration, while data in 2011 and 2012 are applied for model validation. The auto calibration tool of SWAT-CUP (version 5.1.6.2) with its optimization program of SUFI-2 is adopted for the model calibration, validation and uncertainty analysis in this study. The two indices of streamflow and sediments are performed together regarding calibration and validation in SWAT-CUP. The algorithm of SUFI-2 considers all uncertainty factors (e.g. conceptual modelling, input data, parameter non-uniqueness) related in the model on each parameter, which are expressed as uniform distributions or ranges, and tries to capture most of the measured data within the 95% prediction uncertainty (95PPU)[13] of the model in an iterative process (Fereidoon and Koch 2016).

Based on the related literatures (e.g. Abbaspour et al. 2015; Yang et al. 2013a; Arnold et al. 2012), recommended sensitive parameters for both streamflow and sediments are tested through One-factor-At-a-Time analysis and Global Sensitivity with SWAT-CUP for this study. The final identified sensitive parameters for two indices in this study are summarized in Table A.14 in Appendix 2 in the Electronic Supplementary Material. These parameters are further detailed in terms of soil layers, hydrological groups, and land uses during the process of calibration, and this particular information combined with their corresponding calibrated values is shown in Table A.15 in Appendix 2 in the Electronic Supplementary Material.

For the results of calibration and validation for streamflow and sediments, outputs from SWAT-CUP are illustrated with the observed data, the best simulated values, and the 95PPU, as showed in Figure 6.7 and Figure 6.8 respectively. Regarding the model performance, the related statistical indices of P-factor, R-factor, R^2, NSE, and PBIAS for this evaluation for both results of calibration and validation are displayed in Figure 6.7 and Figure 6.8. The value ranges and the recommended values for the standard of model success regarding these statistical indices are described in Table 6.5 based on previous research (Abbaspour et al. 2015; Arnold et al. 2012). The detailed explanation for each of these statistical indices are demonstrated in Appendix 2 in the Electronic Supplementary Material. Regarding SWAT model calibration and validation, as said by Arnold et al. (2012), the most widely applied statistical indices are R^2 and NSE.

According to the standard in Table 6.5, the results in Figure 6.7 and Figure 6.8 are acceptable for this study, although not perfect (P-factor for streamflow in the validation period and for sediments in both calibration and validation period are not stratified). Due to the data limitation, the observed data for calibration and validation are only four years, which is a bit short time period to get good results. Regarding the statistics of R^2 and NSE, the results for sediments are better than that for streamflow for both calibration and validation. This is mainly because

---

[13] The 95PPU is calculated at the 2.5% and 97.5% levels of the cumulative distribution of an output variable obtained through Latin hypercube sampling.

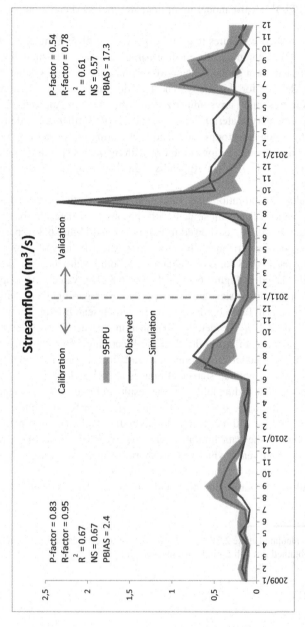

Figure 6.7 Illustration of full SWAT-CUP output for discharge (Source: Own results)

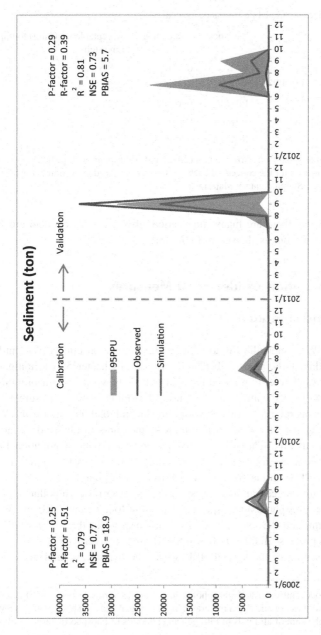

Figure 6.8 Illustration of full SWAT-CUP output for sediment yield (Source: Own results)

Table 6.5 Indices for model performance evaluation

Function	Statistics	Value range	Acceptable standard(monthly time step)
Model accuracy (calibration and validation)	R^2	0 to 1	≥ 0.6
	NSE	$-\infty$ to 1	≥ 0.5
	PBIAS	0 to 100	$\leq 20(\%)$
Uncertainty analysis	P-factor	0 to 1	≥ 0.7
	R-factor	0 to $+\infty$	≤ 1.5

Source: Content adapted from Arnold et al. (2012) and Abbaspour et al. (2015)
Note: The explanation for the concept of each of these statistics is demonstrated in Appendix 2 in the Electronic Supplementary Material

that for sediments there are many time periods that the loads of them are zero due to weather reason, as showed in Figure 6.8.

6.4   Simulation of Identified Measures

6.4.1   General Strategy

The aim of SWAT modelling is to simulate the mitigation effects of identified measures in this study, as described in Table 5.2 in Chapter 5. The mitigation effects of measures refer to reduced loads of pollutants when implementing measures compared to BAU situation (conventional farming without measures). The relevant pollutants refer to sediment, total nitrogen and total phosphorus in SWAT model[14]. The goal of the mitigation effects of measures in this study is at the whole watershed level. Therefore, in SWAT model the data of pollution loads are all gathered at the total outlet of the whole watershed, which refers to the westernmost HRU with the code of 24 as illustrated in Figure 6.6.

Based on this, one certain measure's mitigation effect for each pollutant (sediment, nitrogen, phosphorus) in a special spatial unit (one special HRU in SWAT) is simulated through the representation of this measure in that special HRU in SWAT. In this process, all other HRUs in SWAT are kept as BAU situation, changing only the setting of the special HRU where the measure is implemented for

---

[14] In SWAT model, nitrogen and phosphorus have various forms respectively, like NO3, NH4, NO2, as well as organic P and mineral P. Total nitrogen and total phosphorus in SWAT model refer to the sum of all kinds of nitrogen and all kinds of phosphorus respectively.

the special measure representation. The mitigation effect of the special measure in the special HRU is the reduction amount of each pollutant's load at the total watershed outlet (the load of sediment, total nitrogen and total phosphorus at the outlet with the code 24 under BAU subtracts the corresponding load respectively under measure simulation). In order to obtain the different mitigation effects of a measure in the 50 heterogeneous HRUs covered with cropland, the measure needs to be represented respectively in each of the targeted 50 HRUs. Each of the identified measures needs to be simulated for their respectively heterogeneous mitigation effects in the 50 HRUs. Therefore, the simulation with SWAT should be performed for each of the identified 12 measures in each of the 50 HRUs, as illustrated in Figure 6.9. In total. There are 600 times of simulation needed with SWAT model for the mitigation effects assessment of measures in this study.

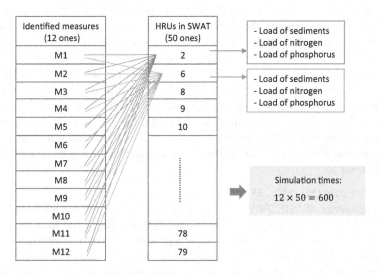

Figure 6.9 General strategy for simulation of identified measures (Source: Own results and drawing, Note: for the 12 identified measures, please check the exact names corresponding to the codes from M1 to M12 in Table 5.2; only the 50 HRUs covered with cropland are considered here, please check the exact locations of them corresponding to the codes from 2 to 79 in Figure 6.6)

As the mitigation effects of measures are targeted at the whole watershed level, the simulation for each measure in each HRU needs to be performed individually, which induces totally 600 times of simulation. By doing this, the aim is to consider the impact of routing and in-stream processes for pollution transport in this study. Many research does the simulation of mitigation effects of measures with SWAT model at the sub-watershed or HRU level (Maringanti et al. 2009; Maringanti et al. 2011; e.g. Arabi et al. 2006), i.e. the pollution load at the outlet of each HRU is compared for the calculation of measure effects. The research has the implication that assumes the influence of in-stream processes is minimal for pollutant transport from HRU outlet to the total outlet of the whole watershed.

6.4.2 Representation of Measures in SWAT

The representation of each identified measure in SWAT model is performed through the adjustment of the values of the appropriate model parameters. For structural measures, it is to identify the corresponding process parameters and then to alter the value of these parameters properly to mimic the functionality of the measures. For operational measures, it mainly refers to the management input file (.mgt), changing the procedures of operations and related values correspondingly in the file based on the measure definition and instruction. There are literatures and guideline reports specially developed for the method of the representation of different kinds of measures in SWAT. In this study, according to Arabi et al. (2008), Waidler et al. (2011), and Arnold et al. (2013), the detailed operations for the representation of each of the 12 identified measures are described as follows according to their category types (as described in Table 5.2 in Chapter 5).

(I) Filter strip (M1, M2, M3)
The parameter of "FILTERW" (in the file.mgt in SWAT) refers to the width of edge-of-field filter strip with the unit of meter. When there is no filter strip (or similar measures like field borders) being represented in SWAT model, the value of this parameter is zero. In this study, under the BAU situation, the value of the parameter of "FILTERW" is zero. For representing the measure of filter strip with different widths of 5 meters (M1), 10 meters (M2) and 15 meters (M3), it is to adjust the value of the parameter of "FILTERW" from zero under BAU to the value of 5 for M1, 10 for M2, and 15 for M3 respectively in the corresponding HRU of the model.

(II) No-till (M4)
No-till, as the operational measure, the representation of it in this study in SWAT is to delete all the activities (in the file.mgt under the category of "Operations")

related to tillage operations under BAU situation, including the plowing and harrowing. At the same time, it needs to adjust the seeding operation in the.mgt file
with the less tillage machine (selecting under the category of "TILL_ID" in the
model).

(III) Nutrient management (M5, M6, M7, M8)
Measures regarding nutrient management have two types in this study, as showed
in Table 5.2 in Chapter 5. One is purely reducing the amount of chemical fertilizer
(in this study it refers to nitro-phosphate) with different degrees of reduction by
25% for M5 and reduction by 40% for M6. The other is reducing the amount of
chemical fertilizer by 50%, combined with applying swine manure with 1000 kg/
ha for M7 and applying sheep manure with 1000 kg/ha for M8. The representation
of M5 and M6 is to change the amount of chemical fertilizer of nitro-phosphate
in the file of.mgt from the amount of 750 kg/ha under BAU to the amount of
562.5 kg/ha and 450 kg/ha respectively for each crop. For measures of M7 and
M8, it is to change the application amount of nitro-phosphate in the .mgt file
from the amount of 750 kg/ha to 375 kg/ha, and at the same time to add an
operation in the.mgt file for manure application with sheep manure and swine
manure respectively with the amount of 1000 kg/ha.

(IV) Cover crops (M9, M10)
Measures of cover crop with soybean and corn (M9 and M10) are to protect
the bare cropland with covers of crop. The representation of them is to set only
the procedure of the seeding of the cover crop in the beginning of June and
the procedure of harvesting of the cover crop in the beginning of September, as
during this period the cropland under BAU in the Baishahe watershed is bare.
However, for some HRUs which have both the winter wheat and corn being
planted in a year under BAU situation (as showed in Figure 6.6), the cropland
from the beginning of June to the beginning of September is not bare but has the
planting of corn with all conventional activities. For these HRUs, the operation
activities for winter wheat are not changed in management file.mgt, while all
activities related to the conventional corn planting are deleted and replaced with
the seeding of cover crop in the beginning of June and the harvest of cover crops
in the beginning of September.

(V) Compounded measures (M11, M12)
Compounded measures in this study are the combination of no-till and reducing
the chemical fertilizer by 25% and by 40% respectively (M11 and M12). The
representation of it would be the combination for the changes of no-till and the
amount of chemical fertilizer reduction described before at the same time. That

is for a special HRU, in the.mgt file, all tillage activities related to ploughing and harrowing are deleted and seeding changed with less tillage machine (no-till planter), at the same time the application amount of nitro-phosphate is changed from 750 kg/ha under BAU to 562.5 kg/ha and 450 kg/ha respectively for M11 and M12.

6.4.3 Results and Analysis

From the SWAT model simulation, the quantified mitigation effects of each measure in each HRU are resulted, as shown in Appendix 6 in the Electronic Supplementary Material. The simulated mitigation effects for each measures along with the SHUs are different. The reasons for the heterogeneous results are various, including the heterogeneity of the locations, elevations, soil types, and land use kinds of the different spatial units. However, along with the same sequence of the SHUs, there are some regular changing trends for the mitigation effects of measures, as illustrated in Figure 6.10, Figure 6.11 and Figure 6.12 for pollutants of sediment, nitrogen and phosphorus respectively. There are six sub-figures in each of these figures for each of the three targeted pollutants (Figure 6.10, Figure 6.11 and Figure 6.12), with each sub-figure refers to the measures under the same category (please refer to Table 5.2 in Chapter 5 for measures). For all these figures, the X-axis represents the SHUs (referring to HRUs) which are sequenced in the same order and distinguished with A-zone and B-zone. A-zone refers to the 16 spatial units which have both wheat and corn planting in the study region (the blue area in Figure 6.6 in Chapter 6), and B-zone refers to the other 34 spatial units which only have wheat planting in the study region (the yellow area in Figure 6.6 in Chapter 6).

Generally, the mitigation effects for the sub-measures under the same kind of measure category have similar changing trend along with the change of SHUs. Different measure categories have totally different situations, while for some measure categories the mitigation effects for nutrient pollutants of nitrogen and phosphorus have similar changing trends along with the SHUs, as showed in Figure 6.11 and Figure 6.12.

(I) Sediment
Regarding the mitigation effects for the pollutant of sediment, there are negative values for all measures except filter strips (M1, M2, and M3), as showed in Figure 6.10. This is because that filter strips are structural measures, which would not impact the growth of the crop plant. However, other measures are operational

practices, which might influence the growth of crop plant more or less and then influence the ability of the crop plant to prevent the loss of sediments. Along with the same sequence of the SHUs, for all measures the charts have high values and low values, eight positive values or negative values. These high and low values for each measure are the corresponding same spatial units, which means that these spatial units relatively have higher sensitivity compared to others regarding the pollutant load of sediment due to their characteristics like locations, elevations, soil types, etc. Negative values in Figure 6.10 for the sediment load reduction means that the measures in the corresponding spatial units are not environmental effective, which cannot help to reduce the sediment load but leads to the increase of the pollutant load.

Measures under the same category have the similar changing trend for sediment load reduction along with the sequence of SHUs for all measures. Especially for M1, M2 and M3, they are filter strips with different width of 5 meters, 10 meters and 15 meters respectively, which have the same changing trends for mitigation effects along with the spatial units. The wider the filter strip is the higher the mitigation effects they have, while the increased mitigation effects are not big as the same magnification as their increase of the width.

Regarding M5 and M6 (chemical fertilizer reduction 25% and 40% respectively), as well as M7 as M8 (chemical fertilizer reduction 50% combined with swine and sheep manure respectively), many of the spatial units for them have negative values, as showed in Figure 6.10-C and -D, while there are still many positive values which is close to zero. This reflects that these two categories of measures (M5 and M6, as well as M7 and M8) are not effective for mitigating sediments, as their operations are to reduce the amount of chemical fertilizer input into the cropland. Fertilizer reduction in some spatial units could induce the poorly growth of crop plants, which have the function of preventing the loss of sediment. When the stem of crop plants are not as strong as before, the function for preventing sediment will weak.

According to Figure 6.10-B and -F, the charts are generally the same. This is because that M11 and M12 (combination of no-till and fertilizer reduction) are the compounded measures based on M4 (no-till) as well as M5 and M6 (fertilizer reduction differently), where M4 have relatively much higher magnitude of values in each spatial unit for sediment load reduction compared to that of M5 and M6. Therefore, measures related to no-till have relatively good mitigation effects for pollutant of sediment.

For M9 and M10 (cover crops), as showed in Figure 6.10-E, it is noticeable that in general the spatial units in A-zone have negative effects, and the spatial units in B-zone have generally positive mitigation effects. This is because that for measures of cover crop (M9, M10) in this study, their operations are performed as there is no-till activities and no fertilizer being applied along with the planting of cover crops. The spatial units in A-zone have corn being planted under BAU at the time period for planting the cover crops under the measure scenarios of M9 and M10. With the implementation of M9 and M10, it means that the original corn under BAU which have tillage activities and fertilizer application has better stem growth conditions than the cover crops without tillage activities and fertilizer application. Therefore, original corn under BAU have better ability to prevent sediment loss, as showed in the Z-zone area in Figure 6.10-E. However, for the spatial units in B-zone, there is no crop planted under BAU during the time period for planting cover crops with M9 and M10. With the implementation of M9 and M10, there are crop plants covered on the land compared to the situation that the cropland is bared under BAU. For preventing the sediment loss, it is much better for cropland covered with plant than cropland which is bared. This is the reason that why the spatial units in B-zone in Figure 6.10-E the values of the reduced amount of sediment load are generally positive.

(II) Nitrogen and phosphorus

Nitrogen and phosphorus are nutrients, which have similar principles for generating pollutant loads. According to the sub-figures in Figure 6.11 and Figure 6.12, the two pollutants have similar changing trends for the pollution load reduction along with the same sequence of the SHUs for some measures, especially for the spatial units in B-zones. These measures include M1, M2 and M3 (filter strips), M4 (no-till), M9 and M10 (cover crops), as well as M11 and M12 (compounded measures of no-till and fertilizer reduction). These measures, like filter strips and others related to no-till, are more for preventing the nutrient loss in other ways instead of reducing the amount of fertilizer application. The general difference for the mitigation effects of nitrogen and phosphorus is that for each measure the mitigation magnitude for nitrogen is bigger than that of phosphorus. This is because that the amount of pollutant load of nitrogen under BAU is much bigger than that of phosphorus in the study region, according to the information of the local situation (J.J. Jin, C. Guo, Water Conservancy Bureau, Xia county, personal communication, January, 2018) and the results from SWAT model simulation. For both nitrogen and phosphorus, the mitigation effects of measures have negative values for some spatial units, except measures of filter strips (M1, M2, and M3) and chemical fertilizer reductions (M5 and M6). The negative values means that

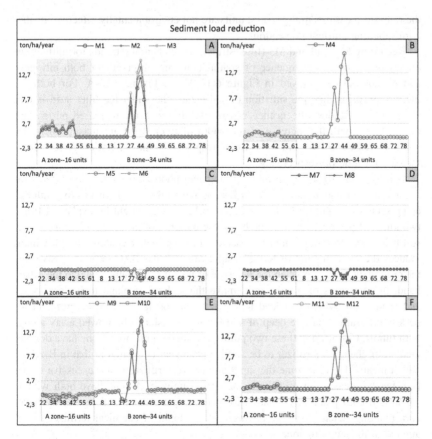

Figure 6.10 Sediment load reduction of measures along with SHUs (Source: Own results, Note: Y-axis is the reduced amount of sediment load of measures, with positive values meaning a reduction of pollutant and negative values meaning an increase of pollutant. X-axis are 50 HRUs, with 16 ones in A zone (cropland planting both corn and wheat under BAU) and 34 ones in B zone (cropland planting only wheat under BAU) (the numbers on X-axis are codes of HRUs, not all of them are shown). There is no functional relationship between Y-axis and X-axis. M1 to M12 represent the twelve identified measures, please refer to Table 5.2 in Chapter 5 for the meanings)

the relevant measures in the spatial units are not environmentally effective for the corresponding nutrient prevention.

Regarding M1, M2 and M3 (filter strips), same changing trends for mitigation effects along with the sequence of spatial units are observed for both nitrogen and phosphorus, as showed in Figure 6.11-A and Figure 6.12-A. For both of the nutrients, just like the situation of sediments, the wider the filter strip is the higher the mitigation effects they can result. However, the increased mitigation effects along with the increase of the width of filter strips are linear, when the width increases to two and three times respectively the mitigation effects only increased a little bit respectively.

For M4 (no-till), for both nitrogen and phosphorus, the mitigation effects in A-zone have negative values and in B-zone have relatively high positive values, as presented in Figure 6.11-B and Figure 6.12-B. This might be explained from two kinds of perspectives. On one hand, the measure of no-till could improve the soil health and in return could increase the organic matter content of soil, which induces the same amount of fertilizer application under BAU to be excessive for the crop plant to absorb when the measure of no-till is implemented. On the other hand, the measures of on-till could make the soil on cropland to be less loose and more hard, which can lead to the nutrients on the surface of cropland to be less infiltrative into the deep of soil but to be easier to be flowed away along with runoff. Considering these two points, the spatial units in A-zone have double amount of chemical fertilizer to be applied each year compared to that in B-zone, which means that in A-zone the input amount of fertilizer is too excessive to be kept by the measure of no-till. As the spatial units in A-zone have both winter wheat and corn planting compared to that the spatial unis in B-zone only have corn planting. Each crop type has the same amount of chemical fertilizer to be applied in the study region.

M5 and M6 refer to reducing the amount of chemical fertilize by 25% and 40% respectively. These two measures can have positive mitigation effects for both nitrogen and phosphorus, as showed in Figure 6.11-C and Figure 6.12-C. For nitrogen, the spatial units in A-zone have much higher mitigation effects than the spatial units in B-zone for both M5 and M6. As A-zone has two kinds of crops, winter wheat and corn, to be planted, and thus the amount of chemical fertilizer reduction is twice than that in B-zone where there is only one crop of corn being planted. M6, chemical fertilizer reduction by 40%, has better mitigation effects in both A-zone and B-zone than M5, which is chemical fertilizer reduction by 25%. However, for phosphorus, although for both measures of M5 and M6 the mitigation effects are both positive, the rules are not regular as the ones for nitrogen.

Regarding M7 and M8, chemical fertilizer reduction by 50% combined with 1000kg/ha animal manure application of swine and sheep respectively, the mitigation effects for nitrogen are basically the same, as showed in Figure 6.11-D. This implies that the swine manure and the sheep manure have minor difference regarding the pollutant of nitrogen generation in the study region. However, as to phosphorus, only M8 has positive mitigation effects generally in all spatial units, while M7 has negative mitigation effects in the majority of the spatial units (Figure 6.12-D). This reflects the big difference for the content of phosphorus in swine manure and sheep manure.

For M9 and M10, M10 (cover crop of corn) has relatively better mitigation effects than that of M9 (cover crop of soybean) for both nitrogen and phosphorus. Soybean belongs to leguminous species, which are commonly considered to have special ability for nitrogen-fixing, and thus are commonly the recommended species for the measures of cover crops (Ordóñez-Fernández et al. 2018; Reckling et al. 2016). Although this, the pollutant load of nitrogen is also influenced by the situation of sediment load, as some chemical forms of nitrogen (mineral N) are attached in sediments to flow into the waterbodies. According to Figure 6.10 for the mitigation effects of sediments, the mitigation effects for M9 is worse than that for M10, especially for the spatial units in A-zone where there are negative values. The situation of sediment mitigation effects influences the situation of mitigation effects of nitrogen and phosphorus. This also explains the negative values of the mitigation effects for M9 and M10 in some of the spatial units, as showed in Figure 6.11-E and Figure 6.12-E.

M11 and M12 are the compounded measures based on the combination of M4 with M5 and M6 respectively. For both nitrogen and phosphorus, the mitigation effects in B-zone for both M11 and M12 have the same changing trend as the situation of M4, referring to Figure 6.11-F and -B as well as Figure 6.12-F and -B. This is because that, for both nitrogen and phosphorus, in B-zone the mitigation effects for M4 have much bigger magnitude than that for M5 and M6, therefore the effects of M4 have predominant functions for the mitigation results of the compounded measures of M11 and M12. Meanwhile, in A-zone, the negative effects for M11 and M12 regarding both nitrogen and phosphorus are less than that for M4, due to the positive mitigation effects in A-zone of M5 and M6.

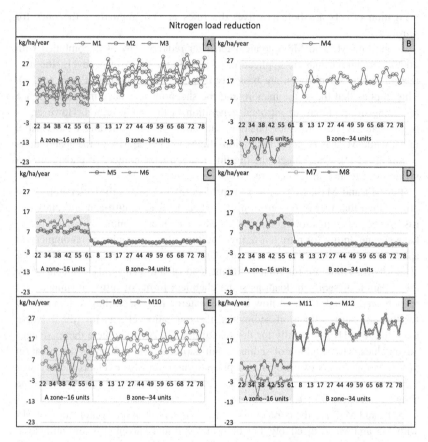

Figure 6.11 Nitrogen load reduction of measures along with SHUs (Source: Own results, Note: Y-axis is the reduced amount of nitrogen load of measures, with positive values meaning a reduction of pollutant and negative values meaning an increase of pollutant. X-axis are 50 HRUs, with 16 ones in A zone (cropland planting both corn and wheat under BAU) and 34 ones in B zone (cropland planting only wheat under BAU) (the numbers on X-axis are codes of HRUs, not all of them are shown). There is no functional relationship between Y-axis and X-axis. M1 to M12 represent the twelve identified measures, please refer to Table 5.2 in Chapter 5 for the meanings)

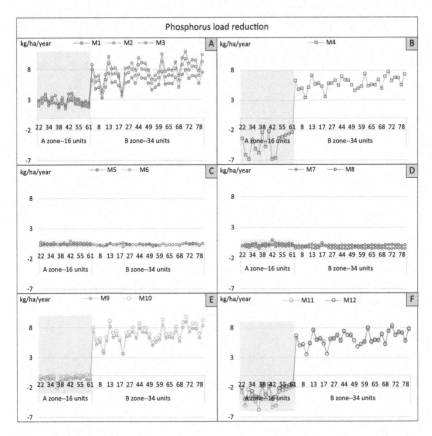

Figure 6.12 Phosphorus load reduction of measures along with SHUs (Source: Own results, Note: Y-axis is the reduced amount of phosphorus load of measures, with positive values meaning a reduction of pollutant and negative values meaning an increase of pollutant. X-axis are 50 HRUs, with 16 ones in A zone (cropland planting both corn and wheat under BAU) and 34 ones in B zone (cropland planting only wheat under BAU) (the numbers on X-axis are codes of HRUs, not all of them are shown). There is no functional relationship between Y-axis and X-axis. M1 to M12 represent the twelve identified measures, please refer to Table 5.2 in Chapter 5 for the meanings)

Agri-Economic Cost Assessment

7

This chapter focuses on the analysis for the cost of mitigation measures. First, the cost category for both ESS buyers and ESS providers in AES programs are described. Second, based on the cost category analysis, the abatement cost for each mitigation measure occurred to farmer when they implementing them are analyzed, and calculated with developed formulas and collected data. Third, the results of the measure abatement costs are analyzed in detail.

7.1 Cost Categories in AES Programs

All kinds of interventions for environmental conservation associate with costs, as there are changed activities due to the interventions (Naidoo et al. 2006). Regarding AES programs, compensation payments are needed from government or other agents to farmers. The amount of compensation payments should be able to cover all kinds of costs that incurred to farmers when they participate in AES programs. To better understand the cost components, the cost categories in AES programs are illustrated in Figure 7.1, where the cost types for both ESS buyers (normally government or society) and ESS providers (normally farmers) are categorized. The explanation for them is below.

Supplementary Information The online version contains supplementary material available at https://doi.org/10.1007/978-3-658-41340-8_7.

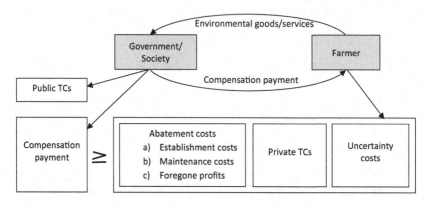

Figure 7.1 Costs involved in AES programs (Source: Modified from Mettepenningen et al. (2009), Note: TCs refer to transaction costs. Abatement costs are also termed as production costs in many literatures, especially for biodiversity conservation (McCann 2013))

7.1.1 Cost for ESS Buyers

For the public agencies who are responsible for organizing AESs, costs are mainly composed of public transaction costs (TCs) and compensation payments for ESS providers (Mettepenningen et al. 2009; McCann 2013; Schöttker and Wätzold 2018). Here the transaction costs happened to ESS buyers are called as the public TCs in order to distinguish the transaction costs happened to ESS providers (private TCs). They might also be described in other terms, such as agency level transaction costs and farm level transaction costs in Schöttker and Wätzold (2018).

Public TCs refer to all the administrative costs caused by operating the schemes, which could be categorized as transaction costs for making decisions and transaction costs for implementing the management decisions (Birner and Wittmer 2004), as described in Table 7.1 with detailed examples. Decision making costs refer to the costs of acquiring necessary information in order to make appropriate decisions, as well as the costs of coordinating between different decision opinions. There is a trade-off between decision making costs and the quality of the decisions. As decision making costs include the costs of getting information for abatement costs and cost-effective allocation of measures. Implementation costs happened mainly to make sure the compliance of conservation activities with environmental legislation and thereby to make sure the success of them, which consist of the monitoring costs and enforcement costs.

Table 7.1   Sub-categories of public TCs

Public TCs			
Decision-making costs		Implementation costs	
information acquiring costs	coordinating decision-making costs	monitoring costs	enforcement costs
e.g., scientific and local knowledge on the effects of measures, preference for conflicting goals, information of production costs	e.g., resources for meetings, settling conflicts, costs due to delayed decisions	e.g., supervisory personnel, specialist equipment	e.g., lawsuits, building and maintaining courts and prisons, collecting fines

Source: Content adapted from Wätzold et al. (2010), Wätzold and Schwerdtner (2005), and Birner and Wittmer (2004)

Compensation payments refer to the monetary transaction between public agencies and farmers based on the transaction condition of the provision of ESS from farmers. As showed in Figure 7.1, the amount of compensation payments should cover at least all the costs occurred to farmers from different perspectives resulting from their participation in AESs.

7.1.2   Cost for ESS Providers

From the perspective of ESS providers, for participating AES programs they would consider all kinds of costs involved and compare the sum of them with the provided compensation payments. These costs generally could divided as three kinds, which are abatement costs, private TCs and uncertainty costs (Figure 7.1). To induce farmers to participate in an AES contract through compensation, the total amount of compensation should cover at least the sum of these costs, that is the participation constraints in AES (Ferraro 2008). Due to the special features and difficulties to assess and quantify the private TCs and uncertainty costs, in this study the focus of cost analysis will be on abatement costs.

(I) Private TCs and uncertainty costs
Private TCs refer to the costs of all kinds of additional activities related to transactions farmers have to perform if they want to participate in the AES program. These activities include gathering information on the program itself and

the offered measures with the program, making decisions on which measures they want to adopt and where, filling out necessary administrative documents, spending time with the program regulators for the compliance with program's monitoring requirements. Private TCs are not trivial, and it has attracted increasing research, such as the sub-categories of TCs are discussed in detail in Mettepenningen et al. (2009).

The uncertainty cost is different as it is not pecuniary cost. It is caused by uncertainties regarding the effects of AES participation on farmers' production level, the effects on the future legal designation of land parcels (e.g. the land parcels might be expropriated by the government), as well as the effects on famers' own social status (e.g. the reputation of a farmer might be affected negatively or positively by other farmers) (Mettepenningen et al. 2009). The uncertainty costs might be subject to an incommensurability problem, as the relevant uncertainties are hard to be recognized and judged. However, these uncertainty factors induce the differences among farmers regarding the attitude of AES participation.

It is hard to assess the values of private TCs and uncertainty costs due to the difficulty to quantify them with clear indices. Methods to simplify the difficulty for solving problems are developed. Regarding common practices in many EU counties, the compensation amount for private TCs is considered as a fixed amount per unit area for all kinds of measures, which is applied also in research, such as Drechsler et al. (2007b), Wätzold et al. (2008). Meanwhile, some studies measure the amount of transaction costs as a proportion of production costs (McCann 2013; Schöttker and Wätzold 2018). For uncertainty costs, it is simplified in Drechsler et al. (2007b) and Wätzold et al. (2008) regarding farmers' attitude towards participation on AES programs through considering a uniform random variable for each farm.

(II) Abatement costs
Abatement costs refer to the costs of the actual activities for carrying out the mitigation measures, which include three types: establishment costs, maintenance costs and foregone profits. These three types of costs are considered to be the major constraints for farmers to adopt the environmental friendly measures (Bekele and Drake 2003; Lapar 1999), and thus are the main consideration for AES compensation payments. In contrast with private TCs and uncertainty costs, abatement costs are relatively easy to be assessed base on the available criteria to quantify the related activities. It would be straightforward to think of foregone profits when considering abatement costs, as foregone profits occur for each kind of measures. However, establishment costs and maintenance costs are

also important components of abatement costs when there are structural measures involved. The main features and differences for these three sub-components of abatement costs are summarized in Table 7.2, with the explanation for each of them below.

Table 7.2 Sub-components of abatement costs

Sub-components	Establishment costs	Maintenance costs	Foregone profits
description	measure installation	keeping expected function	lost net benefits
measure type	structural measures	structural measures	all kinds of measures
happened time	once-off in the beginning	annually	annually

Source: Own analysis

Establishment costs only occur for structural measures, but not for operational measures. Structural measures refer to the land use change practices, such as from cropland to grassland or to forest. For structural measures, the establishment activities for the new permanent land use are needed at the beginning of AES programs and the established new land use will be maintained in the later years during AES lifetime. According to Arabi et al. (2006), establishment costs include costs of the installation of structural measures, and their corresponding technical and field assistance if needed. The characteristic of establishment costs, differing with maintenance costs and foregone profits, is that it is one-time costs only happened at the very beginning of AES programs.

Maintenance costs are accompanied with establishment costs, which means that only structural measures need to be maintained for the expected function each year during the AES lifetime. The maintenance activities refer to all kinds of practices in order to keep the good state of the established new land use. For example, to make sure the expected function of grass filter strip during the AES lifetime, farmers need to check and replant the dead grass, to clean the muddy area, to trim the grass for keeping certain length and then good functions, as well as to take actions to prevent the filter strip to be damaged by vehicles or animals. These activities result in the yearly maintenance costs. The costs of maintenance is usually evaluated as a percentage of establishment cost in research, such as Arabi et al. (2006) and Maringanti et al. (2011).

Foregone profits happen for both structural measures and operational measures. It refers to the reduced amount of the net profit farmers earned after the measure implementation compared to the net profit they earned under the conventional farming situation. Generally, profits earned in conventional farming situation would be more than that under the scenarios of mitigation measure application. As farmers are profit-maximizing pursuers. Foregone profits occur yearly like maintenance costs, which are calculated through the comparison of net profits for farmers between the situation of the profit-maximized farming method (business as usual situation) and the supposed situation with mitigation measures.

7.2   Abatement Cost of Identified Measures

7.2.1   Cost Components for Each Measure

In Chapter 5, twelve mitigation measures are identified for this study. Among these measures, two categories are classified regarding structural measures and operational measures, with each category involves different sub-components of abatement costs, as showed in Table 7.3.

Table 7.3 Abatement costs for different measures

Measures (BMPs)		Abatement costs
Structural practices	M1, M2, M3	Foregone profits + establishment costs + Maintenance costs
Operational practices	M4, M5, M6, M7, M8, M9, M10, M11, M12	Foregone profits

Source: Own analysis
Note: Regarding measures from M1 to M12, please refer to Table 5.2 in Chapter 5

There are three identified measures as structural practices, which have the cost types of foregone profits, establishment costs and maintenance costs as the abatement costs. These three measures refer to the sub-measures of filter strip with different widths. Measures of filter strip, which is to be built on the edge of cropland with the permanent native grass of the study region, requires first to plant the perennial grass strip at the beginning of the AES program and maintain the function the grass strip for the later multiple years. Therefore, the involved abatement costs include the one-off establishment costs of grass planting, maintenance costs yearly and the foregone profits due to the occupied area of cropland yearly.

The left nine measures identified in this study are operational practices, which have only the foregone profits as the abatement costs. These measures do not change the land use permanently, but change the detailed procedures of original land using in certain ways and this change is repeated yearly. Therefore, there are no establishment costs and maintenance costs, while changing the procedures or patterns of conventional land using might change the net profits farmers can earn. For example, regarding measures of reducing the amount of chemical fertilizer, in most cases farmers will lose part of their profits due to the reduction of crop yield.

7.2.2 Average Annual Cost for Each Measure

As demonstrated in Chapter 4, the AES lifetime in this study is five years from 2018 and 2022 and the data for cost calculation of mitigation measures are collected in 2018. As described above, the three cost components (i.e. establishment costs, maintenance costs, foregone profits) happen at different temporal points. However, for the aim of simulation and optimization modelling procedure with mitigation effects and economic costs of measures in this study, the abatement cost of each measure in each heterogeneous spatial unit must to be a value of average annual result. Here, the objective is to get this average annual cost value with that the cost components of establishment costs occur at the beginning of AES program, maintenance costs and foregone profits occur every year during AES program, as well as that the cost data are gathered in the beginning year of AES program.

According to Boardman et al. (2017), when making intertemporal (across time) comparisons for the values of costs or benefits of projects, discounting or compounding must be applied so that all values of compared items are in a common metric. The logic is that it can only compare the values of costs or benefits based on the same time point, due to the time value of money and inflation. Discounting is the method that calculating the present values (PVs) of all the benefits and costs of projects before doing the analysis. On contrast, compounding refers to the method of calculating and comparing the future values of all costs and benefits. In this study the method of discounting is adopted for solving the relevant problem, while results would be the same with the method of compounding. With the method of discounting, first all cost components happened in each year during AES lifetime are discounted into PVs in 2018, second the average annual cost is assumed to be c_a in each year in AES lifetime, which is also discounted into PVs in 2018. These two resulted PVs should be equal and the assumed average annual cost of c_a could be calculated. The process is descried as follows.

(I) Present value of cost components

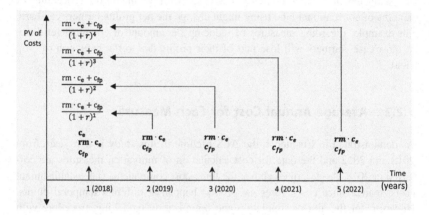

Figure 7.2 Time line of all arisen costs (Source: Own results and drawing, based on the concept of Boardman et al. (2017), Note: c_e: establishment costs; $rm \cdot c_e$: yearly maintenance costs; rm: ratio of maintenance to establishment cost; c_{fp}: yearly foregone profits; r: real interest rate as the discount rate)

As showed in Figure 7.2, the time line of the three cost components according to their arisen time during AES lifetime is illustrated. For abatement costs of measures, in the first year establishment costs (c_e) occur along with the first year's maintenance costs ($rm \cdot c_e$) and foregone profits (c_{fp}), followed with that only the maintenance costs ($rm \cdot c_e$) and foregone profits (c_{fp}) appear each year for the next four years. In Figure 7.2 all three cost components are represented with c_e, $rm \cdot c_e$ and c_{fp} in real terms, without consideration of inflation[1] at each happened time point. Therefore, when discounting the later four years' costs values into PVs, the discount rate needs to be a real interest rate. According to

---

[1] Alternatively, if the values of three cost components are represented in time line in nominal terms, with the consideration of inflation, then we need to do the discounting using nominal discount rate, in which way the final result of summed PVs are the same.

Figure 7.2, the summed PVs for all arisen costs during AES design life could be demonstrated as:

$$PV_{costs} = c_e + \left(rm \cdot c_e + c_{fp}\right) + \frac{rm \cdot c_e + c_{fp}}{(1+r)^1} + \frac{rm \cdot c_e + c_{fp}}{(1+r)^2}$$
$$+ \frac{rm \cdot c_e + c_{fp}}{(1+r)^3} + \frac{rm \cdot c_e + c_{fp}}{(1+r)^4}$$
$$= c_e + \left(rm \cdot c_e + c_{fp}\right) \cdot \sum\nolimits_{i=1}^{5} \frac{1}{(1+r)^{i-1}} \qquad \text{(Eq. 7.1)}$$

where, PV_{costs} indicates the summed present value of the three cost components of the abatement cost during AES program; and for the meanings of related variables of c_e, rm, c_{fp}, and r, please refer to Figure 7.2.

(II) Formula of average annual cost
In order to obtain the average annual value of abatement costs of measures during AES design life, it is assumed to be represented by c_a, as showed in Figure 7.3. Here, the expected c_a also refers to the value in real terms as in the studies of Wätzold et al. (2016) and Mewes et al. (2015). Therefore, the real interest rate is needed for the discounting of c_a. As the average annual cost, c_a appear each year during the AES lifetime as showed in Figure 7.3. Each year's c_a is discounted into their present value in the beginning year of AES program. The discount rate needs also to be the real interested rate as before.

The sum of the present values of these average annual costs (c_a) during AES lifetime showed in Figure 7.3 should be equal to the summed present value of all arisen costs of establishment costs, maintenance costs and foregone profits as illustrated in Figure 7.2.

Based on the graphical demonstration in Figure 7.3, the summed PVs of all years' average annual costs are:

$$PV_{c_a} = c_a + \frac{c_a}{(1+r)^1} + \frac{c_a}{(1+r)^2} + \frac{c_a}{(1+r)^3} + \frac{c_a}{(1+r)^4}$$
$$= c_a \cdot \sum\nolimits_{i=1}^{5} \frac{1}{(1+r)^{i-1}} \qquad \text{(Eq. 7.2)}$$

where, PV_{c_a} indicates the summed present value of average annual costs happened each year during AES lifetime; c_a and r refer to the average annual cost in each year and the real interest rate acted as the discount rate respectively, as showed in Figure 7.3.

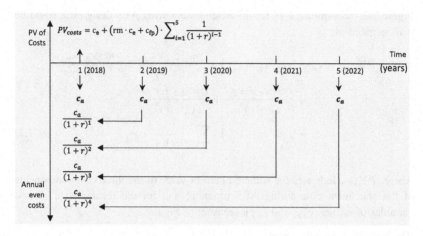

Figure 7.3 Time line of average annual cost during AES design life (Source: Own analysis and drawing, based on the concept of Boardman et al. (2017), Note: c_a: average annual value of abatement costs; r: real interest rate as the discount rate)

As the PVs of average annual costs are the same amount of the PVs of all arisen costs in different time points in Figure 7.2, it has Eq. 7.1 equals to Eq. 7.2:

$$PV_{costs} = PV_{c_a} = c_e + \left(rm \cdot c_e + c_{fp}\right) \cdot \sum_{i=1}^{5} \frac{1}{(1+r)^{i-1}}$$

$$= c_a \cdot \sum_{i=1}^{5} \frac{1}{(1+r)^{i-1}} \qquad \text{(Eq. 7.3)}$$

whereby, the formula for the average annual cost can be resulted as:

$$c_a = c_e \cdot \frac{r \cdot (1+r)^4}{(1+r)^5 - 1} + \left(rm \cdot c_e + c_{fp}\right) \qquad \text{(Eq. 7.4)}$$

Eq. 7.4 demonstrates the general situation for all mitigation measures in this study. However, for the operational measures, with the code of these measures being showed in Table 7.3, the establishment costs and maintenance costs (c_e and $rm \cdot c_e$ respectively in Eq. 7.4) do not exist. Therefore, regarding operational measures with the formula of Eq. 7.4 for obtaining average annual costs, the values of both variables of c_e and $rm \cdot c_e$ in Eq. 7.4 would be zero.

(III) Value of discount rate

According to the above, the value of discount rate should be the value of real interest rate during the AES design life, that is from 2018 to 2022. This is the future time period, the interest rate for this period is based on prediction. The discount rate in this study is aimed for the calculation of abatement costs, which is happened to farmers. The cost assessment is to get the asymmetric information of abatement costs that farmers know but ESS payers do not know in AES programs. Therefore, the cost calculation should be based on farmers' perspective. It is reasonable that farmers might formulate expectation of real interest rate in the future five years based on the average real interest rate happened in the last five years. Based on this, the average real interest rate in the latest five years is adopted as the discount rate for cost calculation in this study, as showed in Table 7.4.

Table 7.4 Average real interest rate in the latest five years in China

Real interest rate (%) in China in the latest 5 years					r
Year	2013	2014	2015	2016	2017	Average
%	3.693	4.732	4.253	3.176	0.284	3.2

Source: Data got from the World Bank (World Bank 2018)
Note: r is the discount rate in Eq. 7.1, Eq. 7.2, Eq. 7.3, and Eq. 7.4

According to Table 7.4, the part involved with discount rate in Eq. 7.4 could be calculated as:

$$\frac{r \cdot (1+r)^4}{(1+r)^5 - 1} \approx 0.2 \tag{Eq. 7.5}$$

7.2.3 Estimation Method of Cost Components

Regarding the analysis and calculation for establishment costs, maintenance costs, and foregone profits for each measure in each heterogeneous spatial unit, this study applies the method of empirical investigation. It refers to that making the estimation of the costs through the local field survey or based on the local officially statistical data and material. The method involves big amount of work of empirical analysis, data collection and detailed calculation.

Abatement costs of mitigation measures refer to the opportunity costs of environmental protection measures. For the assessment methods of the calculation of this opportunity costs, according to Duan et al. (2010), except empirical investigation there are two other methods of direct questionnaire survey and indirect calculation. Method of questionnaire survey refers to obtaining directly the information with questionnaire tables on willingness to pay or willingness to accept from interviewees, such as asking farmers directly the amount of abatement costs for each measure. The disadvantage of it is that the information obtained might be influenced a lot by subjective opinions, which makes the results to have big possibility of uncertainty and inaccuracy. Method of indirect calculation implies that comparing the economic difference of the conservation region with another similar region without conservation to get the opportunity costs of conservation. The similar region is the reference area, which should equip with the very similar natural condition and social-economic development situations with the compared area. However, the method is hard to be feasible and has imaginable weaknesses, making the results of it to have relatively big errors.

With the method of empirical investigation, the costs would be analyzed through the comparison of farming procedures before and after the application of mitigation measures. For doing this, the preliminary work would be to clarify crop production procedures under the business as usual situation and the detailed operations for each measure implementation, in order to get the changed procedures before and after each measure uptake. In this process, the difference of each related factor for the net profit calculation of famers between business as usual situation and conservation scenarios are compared, such as the calculation in Mewes et al. (2015). The related factors involve two main sides, which are the revenue from crop yields and all kinds of costs happened in the crop production process. Therefore, the knowledge of analysis for crop production cost is very important to approach the assessment of abatement costs of cropland mitigation measures. The needed ones for this study are described as follows.

(I) Crop production procedures
The involved crops in this study include winter wheat and corn as the major crops under business as usual situation, and soybean for the measure of cover crop. All these three kinds of crops are cereal, which have the similar crop production procedures. Depending on the economic development, soil, rainfall and others, the cultivation of cereals varies widely in different countries regarding explicit activities, such as the mechanization level in each procedure. However, the general process for crop planting based on the nature of growth is consistent. In China,

the cultivation activities are generally classified as "plowing, harrowing, sowing, management, harvest"[2] corresponding to the timeline of crop growth.

According to the local information (local farmers, personal communication, July, 2018; J.J. Jin, C. Guo, Water Conservancy Bureau, Xia county, personal communication, January, 2018), the cultivation process of cereal under business as usual in the study region could be illustrated as in Figure 7.4, including land preparation, seeding, management and harvest. Here land preparation refers to the work before the sowing of cereal, which are mainly land cleaning for stones and weeds on the field, as well as activities of plowing and harrowing to till and level the soil. Seeding is followed with treated seeds[3]. All the activities between seeding and harvest are categorized as management, including the activities of fertilizing as the starter along with sowing, weeding, pest and disease controlling, irrigating, top-dressing, and general monitoring. At the end, harvesting procedures for cereal crops are mainly reaping, threshing, drying and storage.

Level of mechanization would change the percentages between different cost elements a lot, especially for the labor costs. In the study region, the mechanization level is developed in the past years but the development is limited with the local feature of mountainous terrain. Small machines for plowing, harrowing and seeding are popularized in the area, while big combine-harvesters can only access few cropland patches with good road conditions. Fertilizing for the starter is processed together with seeding and with the same machine, while top-dressing is usually completed manually by farmers. Weeding and pest/disease control are processed with knapsack sprayer manually. Irrigation is only available for patches close to the river without professional machines. Monitoring refers to field checking from time to time by farmers manually, especially after bad weathers like storm. Land cleaning is also done manually with hoes or forks. Regarding harvest, most patches are still finished with labor work for reaping and drying, but application of threshing machines is universal.

(II) Costs categories
There are many different ways to classify the costs of crop production, as showed in Cesaro et al. (2008) and Ciaian et al. (2013). An extensively used way is according to the variation with respect to the unit of production, which classified the costs as fixed costs and variable costs, as showed in Table 7.5. Fixed costs are irrespective of different production levels within certain limits, while variable costs increase with higher levels of production (Cesaro et al. 2008). According

---

[2] In Chinese: "耕、耙、播、管、收".
[3] Sometimes the seeds are treated with pesticide before seeding.

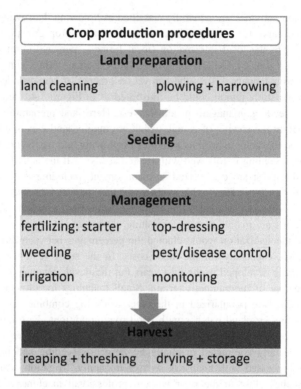

Figure 7.4 Cereal crop production procedures (Source: Own analysis and drawing)

to this, fixed costs include farm management costs, deprecation of purchased machines and land rent costs. Variable costs can be further sub-categorized as costs of material inputs, mechanical operation, manual labor, transport and others, each of which is explained by showing some examples in Table 7.5.

Table 7.5 Typology of production costs

Type of costs	Sub-categories	Examples
Variable costs	Material inputs	Seeds, fertilizers, herbicides, pesticide, fungicide, etc.
	Mechanical operation	Tractor plowing, harrowing, seeding; irrigation facility; harvester, etc.
	Manual labor	Land cleaning, top-dressing, weeding, pest/ disease control, etc.
	Transport	Labor transport, transportation of material inputs and harvested grain, etc.
	Others	Scouting or monitoring, etc.
Fixed costs	Depreciation, land costs, management costs. etc.	

Source: Content adapted from Cesaro et al. (2008), Ciaian et al. (2013), and Iton (2012)

7.3   Cost Calculation and Result

7.3.1   Calculation Formula Development

(I) Changes in revenue and variable costs
For analyzing the cost calculation formulas for each identified measures, first the general formulas for foregone profits, establishment costs, and maintenance costs need to be analyzed. Foregone profits resulted from mitigation measures generally involve changed factors of revenue and all kinds of variable costs, as demonstrated in Mewes et al. (2015). Revenue of farmers is mainly from crop yield. Variable costs refer to the four different kinds of variable crop production costs as described in Table 7.5. In the following, the formulas regarding the changes in revenue and each of these four kinds of variable costs on cropland, corresponding with the activities and items involved in each formula are analyzed, as presented in Table 7.6 with the explanation below.

Changed revenue refers to the changed benefits that farmers earned from the selling of crop yield from the situation of business as usual to the situation of mitigation implementation scenario. Sometimes the crop straw is also considered for cropland benefits, when farmers also sell the crop straw in their conventional situation. However, in the Baishahe watershed the revenue is only from the crop yield, and the crop straw is not effectively used or sold in general. The data for the crop yield under business as usual and under the scenario of each measure implementation in each heterogeneous spatial unit are obtained from SWAT

Table 7.6 General analysis for items in foregone profits calculation

Revenue and cost categories		Activities/items involved	Formulas
revenue	benefit of crop yield	• grain selling	$\sum_{i_1}^{n_1}\left(y_{ref,i_1} - y_{m,i_1}\right) \cdot p_{y,i_1}$ (Eq. 7.6)
variable costs	material input cost	• seeds • fertilizers • herbicide • pesticide/fungicide	$\sum_{i_2}^{n_2}\left(q_{ref,i_2} - q_{m,i_2}\right) \cdot p_{m,i_2}$ (Eq. 7.7)
	mechanical operation cost	• plowing • harrowing • seeding	$\sum_{i_3}^{n_3}\left(o_{ref,i_3} - o_{m,i_3}\right)$ (Eq. 7.8)
	labor costs	• land cleaning • top-dressing • weeding • pest/disease controlling	$\sum_{i_4}^{n_4}\left(l_{ref,i_4} - l_{m,i_4}\right) \cdot p_{l,i_4}$ (Eq. 7.9)
	transport cost	• labor transport	$\dfrac{2D \cdot n_t \cdot s_w}{t_{wd}} \cdot \sum_{i_4}^{n_4}\left(l_{ref,i_4} - l_{m,i_4}\right) \cdot p_{l,i_4}$ (Eq. 7.10)
		• fertilizer and grain transport	$\dfrac{\sum_{i_1}^{n_1}\left(y_{ref,i_1} - y_{m,i_1}\right) + \sum_{i_2}^{n_2}\left(q_{ref,i_2} - q_{m,i_2}\right)}{w_t} \cdot c_t$ (Eq. 7.11)

Source: Own analysis

Note: the meanings of the equations are described in the following context model simulation with the module of plant growth. Based on this, the changed revenue for each crop type is calculated as the amount of crop yield change before and after measure implementation multiplied with the corresponding market price of the grain type. The changed revenue for a heterogeneous spatial unit is the sum of all the changed revenues for all the crop types in that spatial unit. The formula for the revenue change is shown in Eq. 7.6 in Table 7.6, where y_{ref} means the amount of crop yield under business as usual, y_m means the amount of crop yield under measure application, p_y is the market price for the crop grain, i_1 refers to different kinds of crop (i.e. wheat, corn, soybean), and n_1 is the total number of crop types.

The changed material input costs means that before and after measure implementation the input material to the cropland might be changed, which induces to the changed costs. In this study, the considered cropland input material for this cost calculation include four kinds, as showed in Table 7.6, which are seeds, fertilizers, herbicide, and pesticide/fungicide. For calculating the changed material cost before and after a measure implementation for each of the material kinds, it is to multiple the changed amount of material with its corresponding market price. The changed material input costs for a heterogeneous spatial unit is the sum of the changed material cost for all the material kinds. The formula for it is presented in Eq. 7.7 in Table 7.6, where, q_{ref} is the quantity of input material under BAU, q_m is the quantity of input material under measure application, p_m is the market price of input material, i_2 refers to different kinds of materials, and n_2 is the total number of material types.

Regarding mechanical operation costs, in the Baishahe watershed the way for farming activities with machine using (activities of plowing, harrowing, seeding) is trusteeship between agricultural machine owners or operators and farmers. The trusteeship refers to that farmers pay machine operators according to the cropland area for letting them to complete certain activities (i.e. plowing, harrowing and seeding) on the cropland. There are market prices in the local area for each of the farming activities with the mechanical operation based on trusteeship. Based on this situation, instead of considering machine type, service time, fuel price, maintenance costs, and depreciation rate, mechanical operation costs in this study are calculated directly with the cost of mechanical operation based on trusteeship. The changed mechanical operation costs for a heterogeneous spatial unit is the sum of all kinds of the changed mechanical operation costs in that area. The formula for it is presented in Eq. 7.8 in Table 7.6, where, o_{ref} and o_m are mechanical operation costs with trusteeship under BAU and under measure application respectively, i_3 refers to different kinds of farming activates with mechanical operation, and n_3 is the total number of farming activates with mechanical operation.

Labor costs involve activities of land cleaning, top-dressing, weeding, as well as pest and disease control in this study, as these activate are conducted manually by farmers in the study region with necessary small farming tools. The changed labor costs before and after measure implementation for special labor activity is calculated as the changed labor time of the activity multiplying with the corresponding market labor price. The changed labor costs for a heterogeneous spatial unit is the sum of all the changed labor costs for all types of labor activities in that spatial unit. The formula for it is shown in Eq. 7.9 in Table 7.6, where, l_{ref} is the labor time used under BAU, l_m is the labor time used under measure application, p_l is the market labor price, i_4 refers to different kinds of manual labor activities, and n_4 is the total number of manual labor work types.

Regarding transport costs, this study consider two types, which are labor transport costs and material transport costs. Labor transport means that when farmers do manual labor work on cropland they need to go and return between their households and cropland patches either using some kind of vehicles or walking. When using vehicles there are costs related to fuel consumption, maintenance and depreciation of vehicles. When walking, more time is spent on the way, which incurred time costs. For the study region, according to the information collected, the majority of farmers walk for doing their manual works on cropland. Therefore, to simplify the situation, walking is assumed as the consistent way for labor transport in the study region.

To calculate the labor transport costs, first the total time spent for walking between cropland and households on a full-time manual labor day of farmers is analyzed. It is the two times of distance between cropland patches and villages multiplying with the number of trips each day of farmer between cropland and villages, and multiplying again with the speed of walking of farmers. Second, this total walking time (with the unit of hours) on a full-time manual workday could be transferred to a proportion of unit labor time (which is daily) through divided it by the market labor hours for a payment labor day. Third, the resulted proportion (of walking time to total labor hours per day) multiplies with the changed labor costs would be the changed labor transport costs. Based on these steps, the formula of changed labor costs for a spatial unit is shown in Eq. 7.10 in Table 7.6, where D is the distance between cropland patches and correspondingly villages, n_t is the number of trips of farmers on each full-time manual workday, s_w is the average walking speed of farmers, t_{wd} indicates the market labor hours for a payment labor day, and $\sum_{i_4}^{n_4}\left(l_{ref,i_4} - l_{m,i_4}\right)\cdot p_{l,i_4}$ refers to the changed labor costs as described in Eq. 7.9.

Material transport costs occurred when transporting all kinds of input materials from villages to cropland patches, as well as when harvesting the crop and transporting them from the cropland patches to villages. Material transport during crop production usually needs transportation facilities, which induces costs. To better do the cost calculation, two points of simplicity are made regarding material transport costs in this study. First, the distance for material transport is assumed from cropland patches to villages instead of from cropland patches to the final marketplaces. As farmers might buy the input materials for cropland and sell their harvested crop grains in their own villages. Second, it is assumed that only the input material of fertilizer (both chemical fertilizer and manure) and the harvested grain are considered for the calculation of the material transport costs, same as in the study of Iton (2012). The reason is that other input materials (e.g. seeds, herbicide, pesticide/fungicide) require quite small amount of quantity input per unit area, and in the study region the cropland patches are quite small and scattered. Therefore, these materials usually do not need the relatively costly transportation facility.

The changed number of times needed for material transport with transportation facility before and after measure implementation is calculated as the sum of the changed amount of input material and the changed amount of grain yield being divided by the carrying capacity of applied transportation facility. The changed material transport costs are then calculated as the changed number of times of delivery with transportation facilities multiplying with the delivery cost per time. Based on this, the formula for the calculation of changed material transport costs is shown in Eq. 7.11 in Table 7.6, where $\sum_{i_1}^{n_1} \left(y_{ref,i_1} - y_{m,i_1}\right)$ and $\sum_{i_3}^{n_2} \left(q_{ref,i_2} - q_{m,i_2}\right)$ refer to the changed amount of harvested crop yield and input materials (only consider fertilizer and manure here) as described in Eq. 7.6 and Eq. 7.7 respectively, w_t is the carrying capacity of applied transportation facility, and c_t refers to the delivery cost per time.

Here, to simplify the problem of obtaining the distance between cropland patches and villages, the data of distance are analyzed and extracted through geospatial analysis (ArcGIS). As showed in Figure 7.5, the locations of the eight villages in the study region are identified through their corresponding longitudes and latitudes respectively. Afterwards, each cropland patch (with the centroid of each HRU) searches its closest village, which acts as the belonged village of the cropland patch. Based on this, the data of distance of each cropland patch to its corresponding village could be extracted from geospatial analysis (ArcGIS).

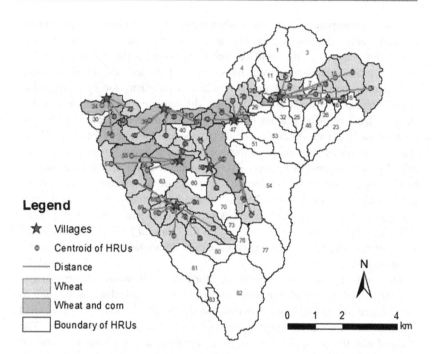

Figure 7.5 Distance from each cropland patch to their corresponding villages (Source: Own result from geospatial analysis, Note: The assessment of the distances is based on the assumption that the village of a cropland patch is the one closest to it on the map, due to the information unavailability for it in reality from the study region)

(II) Foregone profits

Based on the analysis for the change in revenue and four kinds of variable costs before and after measure implementation, the foregone profits of a measure uptake for farmers could be explained. The foregone profits are calculated as that the changed revenue minus all kinds of the changed variable costs. The formula for foregone profits is presented below in Eq. 7.12.

$$
\begin{aligned}
c_{fp} = {} & \sum_{i_1}^{n_1} \left(y_{ref,i_1} - y_{m,i_1} \right) \cdot p_{y,i_1} - \sum_{i_2}^{n_2} \left(q_{ref,i_2} - q_{m,i_2} \right) \cdot p_{m,i_2} \\
& - \sum_{i_4}^{n_3} \left(o_{ref,i_3} - o_{m,i_3} \right) - \sum_{i_4}^{n_4} \left(l_{ref,i_4} - l_{m,i_4} \right) \cdot p_{l,i_4} \\
& - \frac{2D \cdot n_t \cdot s_w}{t_{wd}} \cdot \sum_{i_4}^{n_4} \left(l_{ref,i_4} - l_{m,i_4} \right) \cdot p_{l,i_4}
\end{aligned}
$$

$$- \frac{\sum_{i_1}^{n_1}\left(y_{ref,i_1} - y_{m,i_1}\right) + \sum_{i_2}^{n_2}\left(q_{ref,i_2} - q_{m,i_2}\right)}{w_t} \cdot c_t \qquad \text{(Eq. 7.12)}$$

where, the items of $\sum_{i_1}^{n_1}\left(y_{ref,i_1} - y_{m,i_1}\right) \cdot p_{y,i_1}$, $\sum_{i_2}^{n_2}\left(q_{ref,i_2} - q_{m,i_2}\right) \cdot p_{m,i_2}$, $\sum_{i_3}^{n_3}\left(o_{ref,i_3} - o_{m,i_3}\right)$, $\sum_{i_4}^{n_4}\left(l_{ref,i_4} - l_{m,i_4}\right) \cdot p_{l,i_4}$, $\frac{2D \cdot n_t \cdot s_w}{t_{wd}} \cdot \sum_{i_4}^{n_4}\left(l_{ref,i_4} - l_{m,i_4}\right) \cdot p_{l,i_4}$, and $\frac{\sum_{i_1}^{n_1}\left(y_{ref,i_1} - y_{m,i_1}\right) + \sum_{i_2}^{n_2}\left(q_{ref,i_2} - q_{m,i_2}\right)}{w_t} \cdot c_t$ refer to the changed revenue, changed input material costs, changed mechanical operation costs, changed labor costs, changed labor transport costs, and changed material transport costs respectively. These items are corresponding to the formulas of Eq. 7.6, Eq. 7.7, Eq. 7.8, Eq. 7.9, Eq. 7.10, and Eq. 7.11 in Table 7.6 respectively, with the detailed explanations for each variable in each item being described above regarding changes in revenue and variable costs.

(III) Establishment costs and maintenance costs
Establishment and maintenance costs only involve the measures of filter strip with different widths in this study. Establishment costs are calculated as the summed costs of all kinds of variable costs involved. The category of variable costs for establishment costs are consistent with the category of variable costs analyzed in Table 7.5, including four types of input material costs, mechanical operational costs, labor costs, and transport costs regarding labor transport and material transport. The formula for the establishment costs is shown in Eq. 7.13.

$$c_e = \sum_{i_2}^{n_2} q_{e,i_2} \cdot p_{e,i_2} + \sum_{i_3}^{n_3} o_{e,i_3} + \sum_{i_4}^{n_4} l_{e,i_4} \cdot p_{l,i_4}$$

$$+ \frac{2D \cdot n_t \cdot s_w}{t_{wd}} \cdot \sum_{i_4}^{n_4} l_{e,i_4} \cdot p_{l,i_4} + \frac{\sum_{i_2}^{n_2} q_{e,i_2}}{w_t} \cdot c_t \qquad \text{(Eq. 7.13)}$$

where, c_e is the establishment costs of structural measures. $\sum_{i_2}^{n_2} q_{e,i_2} \cdot p_{e,i_2}$ is the input material costs of measure establishment, where q_e is the quantity of input material for measure establishment, p_e is the market price of input material, i_2 refers to the different kinds of materials, and n_2 is the total number of material types. $\sum_{i_3}^{n_3} o_{e,i_3}$ is the mechanical operation costs of measure establishment, where o_e is the mechanical operation costs with trusteeship, i_3 refers to different kinds of farming activates with mechanical operation, and n_3 is the total number of activates with mechanical operation. $\sum_{i_4}^{n_4} l_{e,i_4} \cdot p_{l,i_4}$ is the labor costs of measure establishment, where l_e is the labor time used during measure establishment, p_l is the market labor price, i_4 refers to different kinds of manual labor activities, and n_4 is the total number of manual labor work types.

$\frac{2D \cdot n_t \cdot s_w}{t_{wd}} \cdot \sum_{i_4}^{n_4} l_{e,i_4} \cdot p_{l,i_4}$ refers to the labor transport cost during measure establishment, where $\frac{2D \cdot n_t \cdot s_w}{t_{wd}}$ indicates the proportion of labor transport time a day to the working time on a full-time labor day of famers, with the detailed explanation for each of the variables in Eq. 7.10 in Table 7.6; and $\sum_{i_4}^{n_4} l_{e,i_4} \cdot p_{l,i_4}$ is the labor costs of measure establishment. $\frac{\sum_{i_2}^{n_2} q_{e,i_2}}{w_t} \cdot c_t$ means the material transport costs during measure establishment, where $\sum_{i_2}^{n_2} q_{e,i_2}$ is the total quantity of input material needed for measure establishment, w_t is the carrying capacity of applied transportation facility, and c_t refers to the delivery cost each time.

Regarding maintenance cost, it is usually evaluated as a percentage of establishment cost in literatures (Arabi et al. 2006; e.g. Arabi et al. 2008; Maringanti et al. 2011). Based on this, the calculation for maintenance costs is shown in Eq. 7.14, where c_n refers to the maintenance costs, rm is the ratio of maintenance costs to establishment costs, c_e is the establishment costs as showed in Eq. 7.13.

$$c_n = rm \cdot c_e \qquad \text{(Eq. 7.14)}$$

(IV) Cost formulas for each identified measure
The analyzed formulas for the cost components of foregone profits, establishment costs, and maintenance costs regarding abatement costs of measures are shown with Eq. 7.12, Eq. 7.13, and Eq. 7.14. These three cost components are the basis for the calculation of abatement costs of mitigation measures, as showed in in the formula of Eq. 7.4 for the average annual abatement costs of measures. All of these formulas are developed based on the general situation of measure implementation, and they are inclusive and comprehensive for all the possible factors. However, for a special measure some considered factors in the general formulas might not be required. Based on these general formulas developed above, the specified formula for each special measure needs to be analyzed.

Regarding each identified measure, the operation procedures for the measure implementation is analyzed and compared with the farming procedures under BAU. In this process, the changed procedures for each measure compared to BAU could be identified. The changed procedures for each measure are the basis for the analysis of considering required factors regarding the calculation on changes in revenue and variable costs, and further regarding the calculation of foregone profits, establishment costs and maintenance costs of the measure. Based on these, the analysis and description for each measure regarding its operations for implementation, changed procedures compared with BAU, as well as the developed specified formula of abatement costs for the measure are presented in Appendix 4 in the Electronic Supplementary Material. The developed formulas for each measure refer to the abatement costs of measures in the SHUs (i.e. HRUs delineated by SWAT model). These SHUs have different shapes and areas, which are considered for the formula development of cost calculation. Regarding operational measures, they are implemented in the whole area of the spatial unit, therefore the area of the spatial unit is considered for the cost calculation. For structural measures of filter strips in this study, they are implemented along with the river of the spatial unit with certain widths, therefore the area constituted by the length of the river in the spatial unit and the width of the filter strip is considered for the cost calculation.

7.3.2 Data Collection

Regarding the data for the calculation of the abatement cost formulas for each identified measure in Appendix 4 in the Electronic Supplementary Material, the primary data collection is needed. The reason is that, first the available secondary data are not sufficient for covering all the data needed for the cost calculation of this study, such as the ones for transport cost calculation. Second, the secondary data are only available for the city where the Baishahe watershed is included. The area of the Baishahe watershed is only a very small proportion of the whole city, thus some of the available secondary data cannot well represent the situation of the local study region. For example, the labor costs can be different in small villages and suburbs of cities. Third, there are no spatial heterogeneity with secondary data. Even the area of the Baishahe watershed is small, the spatial heterogeneity regarding the cost data is expected and only with primary data collection the data heterogeneity could be detected.

Questionnaire survey, as the method of primary data collection, is adopted in this study. The majority of the required data for the cost calculation in this study is based on questionnaire. There are still some data which cannot be obtained through questionnaire survey, like the ones related the operations in measure scenarios that farmers never implement before and have no idea with it. The issue will be solved through other data sources. Besides, some data need to get from the ArcGIS, such as the areas of the SHUs, the length of filter strip (i.e. the length of river) in each spatial unit, and the distances between cropland patches and their belonged villages.

(I) Questionnaire design

According to Puetz (1993), the learning costs and failure rates of household surveys in developing countries are notoriously high, including the unfinished surveys, disappointing data quality, and also the "play-it-safe syndrome", which means collecting more data than actually needed. Sometimes researchers cannot resist the temptation to first collect as enough data as possible and then decide how to use them, which could lead to both resource wastes and no prioritization for the real demanded data. To avoid it, the clear goal of questions, targeted population, sample size, and the method for performing the questionnaire need to be clarified during the questionnaire design. The clarity of these factors could make the survey questions to be formulated as narrowly as possible, and design the questionnaire effectively and practically (Hox and Boeije 2005; Leedy and Ormrod 2005).

The questionnaire is aimed only to get the data for the cost calculation involved in the formulas in Appendix 4 in the Electronic Supplementary Material. However, some related questions are also present in the beginning part of the questionnaire in order to help the targeted population of farmers to warm up about the survey topic. The questions in the questionnaire need to organized in an order that farmers can easily give their answer according to their farming procedures. The survey of questionnaire will be carried out in the way of personal face-to-face and one-on-one interview. The sample size is planned to have five valid households in each of the villages, and then 40 in total with eight villages in the study region. This small sample size is considered to be sufficient for this study because that the general aim of this study is to develop a method instead of giving applicable results. Based on this and due to the time and budget availability for the field survey work (in rural mountainous area with bad road conditions), the sample size is considered to be proper.

During the questionnaire designing, between the first version and the final version of it, efforts were spent on discussion with the experienced experts for questionnaire design as well as pre-tests with the targeted population. After obtaining the preliminary draft of the questionnaire, the pre-tests were done to reveal unanticipated problems, the feedbacks from which are adopted for the discussion with experts and revision. Afterwards, the revised questionnaire went through the pre-tests again. This circled procedure has been repeated several times, as the process showed in Figure 7.6 for questionnaire design. Here, the pre-test was done through telephone interview with the farmers who lived nearby the study region, considering the travelling costs for doing this. The pre-test was performed each time with two farmers, with totally three times of pre-tests being completed in the process.

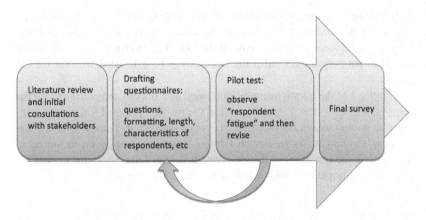

Figure 7.6 Steps of questionnaire design (Source: Modified from Permani (2014))

The final version 'of the questionnaire is completed with three sections, as showed in Appendix 3 in the Electronic Supplementary Material, in both English language and Chinese language. Chapter 1 aims at gaining the basic information about the interviewees and giving simple warm-up questions. The location of the interviewees (village names) is needed for data category, personal information including gender, age and education level would be useful for analyzing the data quality. Questions on the Grain-for-Green Program and other conservation programs are adopted as the warm-up for the theme of mitigation measures in the questionnaire. Besides, questions on cropland lease are necessary, in case the differences of cost situation exist between large-scale farming with leased cropland and the small-household farming.

Chapter 2 is the core of the questionnaire, which includes five parts with different components for relevant crop production costs. Part 1 focuses on the revenue information. Part 2 refers to the transportation costs, with three aspects of human labor transport, harvested grain transport and fertilizer transport (Lucich et al. 2015; Iton 2012). Part 3 aims for the information of farmers' employment cost regarding daily wage and daily working hours. Part 4 focuses on the questions of variable costs during the cropping production. There are two dimensions involved, one is the cropping procedures ordered according to time, the other is the different categories of variable costs. For each cropping procedure, information on each category of the variable costs is questioned. Part 5 targets at the information of farmers' estimation of possible costs for implementing relevant mitigation measure. The hypothetical scenarios of the necessary measures along with their implementation operations are described. Questions are presented with operational procedures of measures and the corresponding variable cost categories for each operational procedure. Not all the identified twelve measure are shown in part 5, but only the ones that the costs cannot be calculated with the data in part 4.

Chapter 3 of the questionnaire is functioned as a smooth and polite end for the questionnaire interview. Farmers are asked to give their personal feeling and opinion after going through the interview. It is open question and has no limits, could be farmers' thoughts for traditional farming, mentioned mitigation measures, or thoughts for the questions in the interview. It is hoped that some unexpected but interesting information could be gained regarding the research.

(II) Survey implementation

The field survey with questionnaire is conducted from the end of June to the end of July in 2018. It is the time period that it is not the busiest period for doing farming work, as well as not the idle period for cropland work when farmers might go to cities for short-term jobs (Knight et al., 2011) or focus busily on family issues. In July it is the time for weeding, a good time period with both the accessibility of famers in the village and their spare time availability for the interview.

Regarding sampling, there are two different techniques, which are probability sampling and non-probability sampling. In this study, the non-probability sampling technique is suitable and adopted, which means that the samples are not selected from a complete sampling frame, thus some individuals have no chance of being selected. Although the technique cannot estimate the very generalizable results compared to probability sampling, it tends to be cheaper, more convenient, and sometimes more practical. There are different methods regarding non-probability sampling, as described and explained in Table 7.7.

The methods of non-probability sampling in Table 7.7 are all applied for the questionnaire interview in this study. First, quota sampling is adopted to divide the population of the whole study region into eight groups, which are the eight villages in the area (the locations of each village is shown in Figure 7.5). For each village, a quota of five interviewees is decided. Second, snowball sampling is applied, that is in each village a resident (the leader of the village) was connected to act as the introducer for the five interviewees in that village. The reason is that the local farmers are very sensitive to the interview and they refuse to do it due to trust issues. At the same time, purposive/judgement sampling and convenience/opportunity sampling are also applied in the process. Regarding purposive/judgement sampling, as the leader of the village know clearly which households

Table 7.7 Non-probability sampling methods

Non-probability (non-random) sampling
Quota sampling: the population is first divided into mutually exclusive sub-group, like stratified sampling; but then judgement is used to select a quota of interviewees from each sub-group
Snowball sampling: when investigating hard-to-reach groups, existing subjects are asked to nominate further subjects known to them, so the sample increases in size like a rolling snowball
Purposive/judgement sampling: relying on the judgement of researcher, people that have expertise or knowledge in the area being researched will be selected
Convenience/opportunity sampling: participants are selected based on availability and willingness

Source: Content adapted from Harrell and Bradley (2009), Permani (2014), and Barratt (2009)

have engaged in cropland work in recent years instead of working in the city as well as which person in the household took charge of the financial situation and could answer the questions well. Along with the judgement and recommendation for the interviewees from the leader of the village, sometimes the wanted interviewees are not at home at the survey time, which makes the convenience/ opportunity sampling to be happened.

Regarding interviews, the method of semi-structured interview is applied. It refers to the extent of control which the interviewer would have over the interaction with the interviewees. Semi-structured interviews have more flexibility compared to structured interviews, and have more control over how the respondent answers compared to unstructured interviews (Wildemuth 2016), as showed in Figure 7.7.

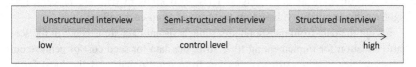

Figure 7.7 Interview types according to control level (Source: Modified from Harrell and Bradley (2009))

With the involvement of the leader of the villages and sometimes other farmers in an interview, researcher asked the questions based on the guide of the questionnaire, while the order of the questions might be changed depending on the talking atmosphere. During the interview process, the village leader and other farmers might discuss and explain the questions in depth with the interviewees. However, they are told in previous to avoid talking in a way that their opinions would influence interviewees' answers. In this way, the interviewees talked more than only the asked questions but also other related farming activities and items, which made the communication spontaneously. This helped to improve the quality of interviews, as Puetz (1993) demonstrates that a congenial atmosphere, patience and good communication with language skills are invaluable assets for an interview.

(III) Data processing
All the data needed for costs calculation are presented in Appendix 4 in the Electronic Supplementary Material. From the questionnaire, data are processed in two kinds of ways depending on the data category. Most of the data categories from the questionnaire are processed with the method of direct average based on all the respondents from all villages. These data are observed to be quite consistent among farmers and villages for each cost category, like the market price of grain, seed costs of each crop, mechanical operation costs, weeding times, etc. It is because that these data are mainly decided by market and the area of the study region is too small to make difference for them.

However, two kinds of data from the questionnaire are detected to have heterogeneity among the eight villages, which are the transport costs for materials with tractors, and the labor costs. Regarding transport costs with vehicles, the heterogeneity is due to the different road conditions in different villages in the mountainous area. For labor costs, it is because that the eight villages have different remote levels, which induces different degrees of labor resources. According to the leaders of each village, they have regulations for the labor price each year, which are different among villages. These two kinds of data are processed as eight results respectively, based on the average from each village.

Except the data from questionnaire, some assistant data are needed. In the scenario situation for implementing filter strip, the data for seed cost of pennisetum (grass species of filter strip) cannot get from questionnaire, as the local farmers never plant this grass before. The data is adopted as the average calculation result, based on the information from online stores in China (the top ten online stores according to the sale amount) for the instruction of seeding rate and the price of pennisetum seed. In addition, some data involved in cost calculation need to be obtained from geospatial analysis and SWAT model, including the area of each spatial unit, the lengths of rivers in each spatial unit for calculating the area of filter strip, the transport distances between cropland patches and villages, as well as the heterogeneous crop yield in spatial units. These four kinds of data are heterogeneous, with the former three being showed in Table A.16 in Appendix 4 in the Electronic Supplementary Material and crop yield data being showed in Appendix 6 in the Electronic Supplementary Material.

7.3.3 Results and Analysis

(I) Negative cost values
Based on all above, the values of costs for each measure in each SHUs can be calculated, with the detailed results showed in Appendix 6 in the Electronic Supplementary Material. The negative cost values are noticed among the resulted cost values, which means that it is profitable for farmers to implement the corresponding measures in the corresponding spatial units even there is no payments for the measures. There are totally four negative cost values, which are all regarding to the measure of chemical fertilizer reduction by 25% (M5) and are in the SHUs with the codes of 38, 41, 58, and 61 respectively in SWAT. These four spatial units have the common feature of that they are all the kind of cropland with the farming pattern of planting winter wheat and corn in turn in each yearly under the BAU situation (please refer to the blue area in Figure 6.6 in Chapter 6). To better explain, the 16 spatial units which have both wheat and corn planting are termed as A-zone in this study (blue area in Figure 6.6 in Chapter 6), and the other 34 spatial units which only have wheat planting are termed as B-zone in this study (yellow area in Figure 6.6 in Chapter 6).

First, the reasons for the negative cost values of these spatial units are analyzed from the perspective of the calculation process with all involved cost components. Regarding the measure of chemical fertilizer reduction by 25% (M5), the abatement costs refer to the foregone profits which involves the cost calculation components of revenue loss, fertilizer input cost change, and the material transport cost change for changed input fertilizer and changed yield. For each of these three cost components, for the spatial units in A-zone both wheat and corn are considered to do the calculation, while in the spatial units in B-zone only kind of crop, wheat, is considered for the calculation. Regarding the cost components of the changed fertilizer input costs and the changed material transport costs, generally the cost values for the calculation in A-zone is the double amount for the cost values in B-zone. However, regarding the revenue loss, the data of the changed grain yield for the cost calculation are obtained from SWAT model, which for the four spatial units (codes of 38, 41, 58, and 61) are relatively lower compared to other spatial units. The small amount of reduced yield induces to the small amount of changed revenue, which is less than the saved costs due to the reduced fertilizer and the reduced material transport costs for fertilizer and yield. This leads to the negative cost values of the measure in these four spatial units.

Regarding the crop yield simulated from SWAT, the nutrients and water resources set in spatial units in SWAT impact strongly the results of crop yield. For spatial units in A-zone, there are irrigation in the model setting, which means these spatial units have relatively better conditions of water resources compared to that in B-zone. When the chemical fertilizer reduced only a little bit (as the operation in M5), the corresponding crop yield will reduce much less for spatial units in A-zone compared to that of spatial units in B-zone without irrigation. For the spatial units with codes of 38, 41, 58, and 61, might due to their special locations in relatively to rivers and other reasons they are just simulated from SWAT model with relatively lower reduced yield for both wheat and corn compared to other spatial units regarding the measure of M5.

Second, the negative cost values of M5 could make sense in general considering the real situations in China. It is well known that chemical fertilizer is over used in China. There is research which makes some recommendations regarding the optimal amount of chemical fertilizer application to maintain relative grain yield and profitability in certain regions of China, like for northern China (Zhang et al. 2018; Xu et al. 2014; Ju et al. 2007). The recommended amount of fertilizer from the research is much less than the amount which is applied under BAU in the study region (which is located in the northern China). This could explain why the proper reduction of the amount of chemical fertilizer in some areas could make farmers to get actually a little bit more net profits. Regarding the reasons why it is more profitable but farmers did not do it in reality, it could be explained with the factors of farmers' traditional habits which are hard to change, unscientific but fixed perception regarding the fertilizer application for them, and poor agricultural extension services for farmers in China (Sun et al. 2012).

Although these negative cost values for the measure in some spatial units are reasonable, considering the followed work of the simulation and optimization for AES design with computer program, the negative values of costs are not feasible. To solve this, the lowest positive cost values of measures (also happened for M5 of chemical fertilizer reduction by 25%) in unit area in this study are adopted to calculate the costs of M5 in the four spatial units which have originally negative values (for the detail please refer to Table A.22 in Appendix 6 in the Electronic Supplementary Material).

(II) Comparison of costs among measures
Based on the values of the costs for each identified measure in each heterogeneous spatial unit (Appendix 6 in the Electronic Supplementary Material), the figures to illustrate the distribution situations of these costs for doing the general comparison between measures are presented in Figure 7.8. There are six sub-figures in Figure 7.8, with each refers to the measures under the same category (Table 5.2 in Chapter 5). For all these figures, the X-axis represents the spatial units which are sequenced in the same order and distinguished with A-zone and B-zone.

In general, the measures of filter strips (M1, M2, M3) and chemical fertilizer reduction by 25% (M5) have relatively lower costs compared to others; while measures of covers crops (M9, M10) have relatively higher costs. For each of the measures the costs are heterogeneous along with the SHUs. Under each of the same measure category the sub-measures have similar changing trend regarding costs along with the sequence of spatial units. The cost heterogeneity is mainly resulted from the factors with heterogeneous values for the cost calculation, including changed crop yield (as showed in Appendix 6 in the Electronic Supplementary Material), as well as distance between cropland patches and villages, length of rivers, labor prices, and transportation factors (as showed in appendix 5 in the Electronic Supplementary Material). Each spatial unit has their fixed values regarding these cost calculation factors, therefore the similar changing trend of costs along with the spatial units under each measure category is reasonable.

Measures of filter strips (M1, M2 and M3) have relatively lower costs in general compared to other measures in this study. It is mainly because that in a certain spatial unit the filter strip is only established in a very small portion of the whole spatial unit. However, the cost per ha for filter strip showed in Figure 7.8-A is calculated as the total cost of filter strip in a spatial unit dividing by the whole area of the spatial unit. Measure of chemical fertilizer reduction by 25% (M5) also has relatively low costs in general, as showed in Figure 7.8-C, that is due to the over use of fertilizer in the study region, as analyzed above for the negative costs values related to this measure. Regarding measures of cover crops (M9 and M10), the relatively high abatement costs in general, as showed in Figure 7.8-E, is caused by both the factors of high reduced yield and high costs for planting cover crops. However, for cover crops, the situation in A-zone and B-zone are opposite. In A-zone, as there is corn planting under BAU at the time period of cover crops, the cover crops (either soybean or corn) with no-till and no fertilizer application induces the very high reduced yield but small amount of saved crop production costs, which results in very high foregone profits. On contrast, in B-zone, as there is no crop planting under BAU during the period of cover crops, the cover crops (either soybean or corn) with no-till and no fertilizer application induces small amount of increased yield but big amount of crop production costs, which results in high foregone profits also.

Regarding the measure category of filter strip, with different widths as sub-measures of M1, M2, M3, they have the same cost changing trends along with the same sequence of spatial units, as showed in Figure 7.8-A. Besides, with the width of filter strip increased from 5 meters (M1) to 10 meters (M2) and then to 15 meters (M3), the cost values corresponding to each spatial unit increase to twice and then three times. The abatement costs for measures of filter strips include foregone profits, establishment costs and maintenance costs. Regarding foregone profits, for filter strips it is the lost benefits when farmers abandon farming totally in a certain area of cropland, which is a constant value in unit area for a certain spatial unit. Regarding establishment costs and maintenance costs of filter strips, they also refer to some fixed values in unit area respectively for a certain spatial unit. Based on these, the abatement costs of filter strips in unit area in a certain spatial unit are consistent. The filter strips of 5 meters (M1), 10 meters (M2), and 15 meters (M3) in a certain spatial unit have different areas. As the length of the river in a certain spatial unit will not change, the area of filter strip with width of 10 meters and 15 meters are two times and three times respectively of the area of the filter strip with width of 5 meters. Therefore, the costs of M1, M2, and M3 have linear increased relationship in each spatial unit. This linear cost relationship only happens for sub-measures of filter strip in this study. For other sub-measures under the same measure category, there are no such linear relationship for costs.

The obvious distinctions between spatial units in A-zone and in B-zone are noticed for all measures except M1, M2, M3 (filters trips with different widths) and M9 (cover crop of soybean), with costs in A-zone are generally much higher than that in B-zone. The distinctions are caused by the different planting situation in A-zone and B-zone under BAU. Regarding measures of filter strips, there is no distinction in A-zone and B-zone due to their establishment costs and main-tenance costs have no relationship to the crop planting situation under BAU, and these two kinds of costs are the main proportion of abatement costs of the mea-sures. Although foregone profits for measures of filter strips relate to the crop planting situation, while the amount of it in general is very small according to

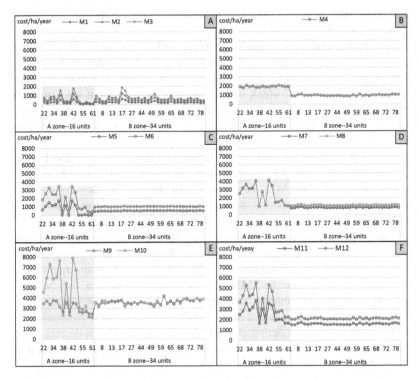

Figure 7.8 Abatement costs of measures along with SHUs (Source: Own results, Note: Y-axis is the costs per ha in RMB of measures. X-axis are 50 HRUs, with 16 ones in A zone (cropland planting both corn and wheat under BAU) and 34 ones in B zone (cropland planting only wheat under BAU) (the numbers on X-axis are codes of HRUs, not all of them are shown). There is no functional relationship between Y-axis and X-axis. M1 to M12 represent the identified measures in this study, please refer to Table 5.2 in 5 Chapter for the meanings)

the calculation in this study, with some spatial units even having negative foregone profits. This reflects that the net profits from cropland for farmers in the study region are very low under BAU, which is also reflected from the questionnaire regarding interviewees' personal opinions for farming. M9 does not show big distinction for abatement cost levels between A-zone and B-zone, which is mainly due to two points. One is that for many spatial units in A-zone, M9 has relatively less amount of revenue reduction compared to M10 in A-zone. The other is that soybean (M9) has relatively lower cropping cost than corn (original crop under BAU in A-zone) in terms of seed input costs, herbicide costs, labor costs, harvest costs and others.

For M4, M5 and M6, M7 and M8, M10, as well as M11 and M12, the abatement costs level in A-zone is much higher in general than that in B-zone, as showed in Figure 7.8. This is because that the spatial units in A-zone have both corn and wheat planting, which induces that the foregone profit calculation in A-zone are the sum of the foregone profits for each of the two kinds of crops. However, the spatial units in B-zone have only wheat planting, with the foregone profit calculation only involving one kind of crop.

Among these measures with obvious distinctions between A-zone and B-zone for abatement costs, M4 has its own patter, while others (M5 and M6, M7 and M8, M10, and M11 and M12) have similar cost changing trends along with the sequence of the SHUs in both A-zone and B-zone. This is because the dominant cost components during the abatement costs calculation for these measures are different. The dominant cost components for M4 are herbicide input costs and labor costs. For other measures (M5 and M6: chemical fertilizer reduction, M7 and M8: chemical fertilizer reduction combined with animal manure application, M10: cover crop of corn, and M11 and M12: no-till combined with M5 and M6 respectively), the dominant component during abatement cost calculation is mainly the revenue change due to the yield change. Besides, these measures have the common feature of reducing chemical fertilizer with different degrees (even for measures of cover crops, it is operated along with no fertilizer application in this study).

Simulation and Optimization of AES

8

This chapter gives the principles of the simulation and the optimization of AES for getting the cost-effective AES design. First, simulation of AES, i.e. farmers' selection behavior in AES, is described along with its assumption, principles, as well as the mitigation effects and summed payments of the resulted land use pattern. Second, optimization of AES is demonstrated, based on the simulation of AES, for its principles and adopted optimization method. Third, the optimization modelling procedure for this study which combines both the simulation of AES and the optimization of AES is presented.

8.1 Simulation of AES

8.1.1 Assumption for Simulation

Regarding the simulation of AES, it is to mimic farmers' voluntary decisions in AES programs on which measures they would like to select for implementation and in which spatial unit. To do the simulation, there is a basic assumption, that is farmers are the pursuers of maximum profits. Farmers will implicitly compare the net economic benefits they could get from the implementation of each of the offered measures in an AES program. They will select for a special cropland patch a measure, among all available measures, which has the maximum positive net economic benefit. If all the available measures have negative net economic

Supplementary Information The online version contains supplementary material available at https://doi.org/10.1007/978-3-658-41340-8_8.

benefits, then no measure will be selected by farmers and they will keep the business as usual farming situation.

Here the net economic benefit for a measure in a heterogeneous spatial unit refers to the result of the payment of the measure in an AES program subtracting the abatement cost of the measure in that spatial unit. The formula for its calculation is shown in Eq. 8.1.

$$NB_{m_i h_j} = P_{m_i} - C_{m_i h_j} \qquad \text{(Eq. 8.1)}$$

where, $NB_{m_i h_j}$ refers to the net economic benefit of farmers for implementing a measure in a spatial unit (HRU in SWAT); P_{m_i} is the amount of payment for the measure in an AES program; $C_{m_i h_j}$ means the abatement cost of the measure in the corresponding spatial unit[1]; m_i refers to a special measure[2]; h_j refers to a heterogeneous spatial unit[3]; i and j refer to the code of different measures and SHUs respectively.

8.1.2 Principles of Simulation

For a special measure, the amount of payment in an AES design in this study is fixed and consistent for all SHUs. However, the abatement costs of a measure in SHUs are distinguished. This induces the net economic benefits of farmers regarding a measure in different spatial units to be heterogeneous, based on Eq. 8.1. There are twelve measures in this study, each of which have heterogeneous net economic benefits for each of the farmers in the 50 targeted spatial units. This leads to the high amount of calculation for the simulation of the selection behaviors of farmers, and computer calculation is needed. To explain the principles during the simulation, a simple example with only three measures (m1, m2, m3) and five SHUs (h1, h2, h3, h4, h5) is given here, with the data for abatement costs and payments being randomly made up by the author and having no relation to the real data in this study. The example is shown in Figure 8.1 and Figure 8.2 with the explanation below.

---

[1] Please refer to Appendix 4 in the Electronic Supplementary Material for the calculation of the abatement cost of each identified measure.

[2] Please refer to Table 5.2 in Chapter 5 for each of the identified measures.

[3] Please refer to Figure 6.6 in Chapter 6 for the 50 targeted SHUs.

(I) List of spatial units for each measure

As showed in Figure 8.1-A, the abatement costs for each measure in each hetero-geneous spatial unit is randomly given in the example. Regarding a measure in all SHUs, the spatial unit which has the maximum net economic benefit will be first selected by farmers to implement the measure. As the payment for a measure is consistent for all SHUs, based on Eq. 8.1 the spatial unit which will have the maximum net economic benefit is the one which has the minimum abatement cost for the measure. For each measure, a list of the SHUs could be generated according to the sequence of the abatement costs from lowest to the highest. This is actually the list of ranked net economic benefits of farmers (Eq. 8.1) from the biggest to the smallest. As showed in Figure 8.1-B, the list is generated for each measure respectively with the abatement cost ranking from lowest to largest. The list represents that, for a given payment for each measure respectively, the net economic benefit of farmers in each of the five spatial unit is ranked from largest to smallest. The list shows the priority of selecting spatial units for implementing a measure from farmers' perspective, and the sequence of the list will not change along with the change of the payments for measures.

A Measures	Spatial units and corresponding abatement costs per unit area				
m1	h1 → 10	h2 → 14	h3 → 5	h4 → 8	h5 → 12
m2	h1 → 25	h2 → 15	h3 → 18	h4 → 20	h5 → 9
m3	h1 → 16	h2 → 28	h3 → 22	h4 → 26	h5 → 33

Note: the values of the abatement costs given to each spatial unit for corresponding measure are randomly given by author and have no relationship to the real abatement costs of this study.

B Measures	List of selection priority of spatial units for farmers				
m1	h3 → 5	h4 → 8	h1 → 10	h5 → 12	h2 → 14
m2	h5 → 9	h2 → 15	h3 → 18	h4 → 20	h1 → 25
m3	h1 → 16	h3 → 22	h4 → 26	h2 → 28	h5 → 33

Figure 8.1   An example to illustrate the principle of list generation for each measure (Source: Based on the principles of the optimization modelling procedure[4], Note: m1, m2, m3 represent three measures, h1, h2, h3, h4, h5 refer to five SHUs in the example)

---

[4] The optimization modelling procedure in this study is programmed by a computer scientist, Dr. Astrid Sturm.

(II) Measure selection for each spatial unit

For selecting each spatial unit with at least one measure, based on the lists generated above, the first ranked spatial units in each list are considered first. For doing it, the payment of each measure must be given. With the given payment for each measure, the net economic benefits of farmers in each of these first ranked spatial units in the lists are calculated and compared. Among these, the maximum net economic benefits would be identified, resulting with the corresponding measure being attached to the corresponding spatial. After this round, the already attached spatial unit with a measure in the corresponding list would be ignored, and the former second spatial unit in the same list would be acting as the first ranked one in this list, so as to repeat the previous procedure for the second round. This process will be repeated and continue until either a measure is attached to each of the spatial units, or the budget for an AES program is used out, or the net economic benefits of all left spatial units are negative.

To better explain the measure selection process, the graphical illustration is shown in Figure 8.2, with the continuation of the previous made up example presented in Figure 8.1. For the three measures, m1, m2, m3, the amounts of payment are given randomly by the author (Figure 8.2-A). Based on the values of these payments, the net economic benefits of farmers for each measure in each of the first ranked cells in the lists are calculated, as showed in the shape of circle with blue color in Figure 8.2-B. Through comparison, the maximum value of 14 is identified, which represents that m3 in h1 has the biggest net economic benefits among the first ranked spatial units, resulting that m3 is attached to h1 as showed in Figure 8.2-C. This is the first round. To continue the second round, the value of 14 in the shape of circle is ignored and the linked arrow leads to the value resulted from the second ranked spatial unit in the same list. This new value is compared with the left other values in the first round to act as the second round (rhombus shape, peach color), which results in that m2 is attached to h5 in Figure 8.2-C. With the same principle, m1 is attached to h3 in the third round (heart shape). In the fourth round (heptagon shape), the maximum value is invalid as the corresponding cell is already occupied in the previous rounds, thus this value is ignored and leads to the next round with the linked arrow. This process continues until the seventh round with all cells are all attached with measures (Figure 8.2-C).

When considering the budget limitation of an AES program, the measure selecting process for spatial units described above must stop at some point. As showed in Figure 8.2-C for the summed payments according to the sequence of the measure attachment process for the spatial units, the summed payments increase along with more spatial units being attached with measures. The measure selecting process will stop at the point when the given budget for the AES program cannot cover the summed payments if the next spatial unit was given a measure.

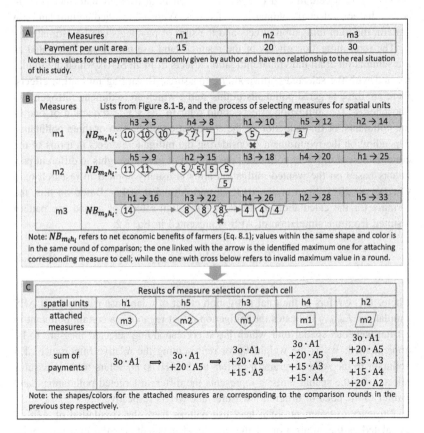

Figure 8.2 An example to illustrate the principle of measure selection for each spatial unit (Source: Based on the principles of the optimization modelling procedure in this study, Note: m1, m2, m3 represent three measures; h1, h2, h3, h4, h5 and A1, A2, A3, A4, A5 refer to the five SHUs and their corresponding areas respectively in the example)

8.1.3   Mitigation Effects and Compensation Payments of AES

Based on the above, under a budget limitation of an AES program and with the payment for each measure being given, the simulation of an AES could be resulted with a certain land use pattern, which is distributed with measures in SHUs that are selected voluntarily by farmers. With the resulted certain land use pattern, the total mitigation effects and the sum of payments can be estimated.

The total mitigation effects of the resulted land use pattern are calculated as the summation of all individual mitigation effects from all SHUs which are implemented with the selected corresponding measures. Regarding mitigation effects, there are three targeted pollutants in this study, which are sediment, nitrogen and phosphorus. The mitigation effects could refer to the single reduction of one of the pollutants, or the reduction of the combination of the two or three pollutants, depending on the requirement. Considering the mitigation effects in terms of the combination of different pollutants, the method of giving weights to different pollutants based on the wanted mitigation aims is usually applied in research (e.g. Chen et al. 2015; Shen et al. 2013; Qi and Altinakar 2011). Based on this, the formula for the calculation of the total mitigation effects of a land use pattern resulted from the simulation of AES is presented in Eq. 8.2.

$$E_{AES} = \sum_{h_j g} W_g \cdot E_{m_i h_j g} \qquad \text{(Eq. 8.2)}$$

where, E_{AES} refers to the total mitigation effect of a land use pattern from AES simulation; $E_{m_i h_j g}$ indicates the mitigation effect regarding a certain pollutant in a heterogeneous spatial unit which has a corresponding selected measure to be implemented; W_g is the weight factor representing the relative importance of a pollutant in the AES program; h_j refers to the SHUs; m_i refers to the measures selected for the spatial units; g represents the different targeted pollutants; i and j refer to the code of different measures and SHUs respectively.

The sum of payments of a land use pattern from the simulation of AES is calculated as the summation of the individual payments for the measure implementation in each of the SHUs. In this study, the area of the SHUs is not consistent, and the information of payment for each measure in an AES is given in terms of the amount per unit area. Therefore, the payment for a spatial unit with a certain measure implementation is calculated as the amount of payment for the measure per unit area multiplying with the area of the corresponding spatial unit, as showed in Figure 8.2-C with the made up example. Based on this, the formula for the calculation of the sum of payment of a special land use pattern resulted from the simulation of AES is shown in Eq. 8.3.

$$SB_{AES} = \sum_{h_j} P_{m_i} \cdot A_{h_j}$$
(Eq. 8.3)

where, SB_{AES} refers to the sum of budget of a land use pattern from AES simulation; P_{m_i} indicates the amount of payment per unit area for a measure in an AES program; A_{h_j} refers to the area of a heterogeneous spatial unit; h_j refers to the SHUs; m_i refers to the measures selected for the spatial units; i and j refer to the code of different measures and SHUs respectively.

8.2 Optimization of AES

8.2.1 Principles of Optimization

Based on the simulation of AES, the optimization of AES could be processed for getting the cost-effective AES design with the land use pattern that is resulted from farmers' voluntary selection for measures. Regarding cost-effective AES design, there are two alternatives. One is to achieve the maximum mitigation effect under the given budget limitation, and the other is to reach the minimum budget aim under the condition of obtaining the required level of mitigation effects. In this study, for cost-effective AES design it will only focus on the former alternative of achieving the maximum mitigation effects under the given budget limitation.

As showed in Figure 8.2, with the given payment for each measure and the given budget of the AES program, a land use pattern is got from the simulation of AES. In the process, the amount of the payment for each measure is very important, as with different sets of payments for measures there will be different land use patterns resulted along with different total mitigation effects. The optimization of AES is achieved through the process of changing randomly and iteratively the amount of payment for each measure in the procedure of simulation of AES (refer to Figure 8.2-A). The aim of it is to compare and find which set of payments for measures could get the land use pattern which has the maximum total mitigation effects (refer to Eq. 8.2) under a certain budget limitation. Based on the description for the total mitigation effects and the sum of payments of a land use pattern from the simulation of AES, the objective function for the optimization of AES is shown in Eq. 8.4.

$$E_{AES} = \sum_{h_j g} W_g \cdot E_{m_i h_j g} \rightarrow \max, \text{ subject to } SB_{AES} = \sum_{h_j} P_{m_i} \cdot A_{h_j} \leq B_0$$

(Eq. 8.4)

where $E_{AES} = \sum_{h_j g} W_g \cdot E_{m_i h_j g}$ is from Eq. 8.2 representing the value of the total

mitigation effects of an AES design with a land use pattern; $SB_{AES} = \sum_{h_j} P_{m_i} \cdot A_{h_j}$

is from Eq. 8.3 representing the value of the sum of payments of an AES design with a land use pattern; B_0 refers to the value of the given budget for an AES design.

The operation of the optimization of AES needs to be conducted through an appropriate optimization method for this study. The iteration and comparison process of optimization is complex due to the problem of local optimization trap. To avoid the issue of local optimization trap, in this study the optimization method of simulated annealing is adopted, which is capable to get at least the approximate global optimal result within a limited number of times of iteration and thus within a limited period of time.

8.2.2   Overview of Simulated Annealing

Simulated annealing (SA) was developed by Kirkpatrick et al. (1983) for solving large combinatorial optimization problems, which is based on the Metropolis Monte Carlo algorithm (Metropolis et al. 1953). As reflected in the name of SA, it is inspired by the physical process of metal annealing or cooling in metallurgy. In condensed matter physics, when heating up a solid metal to enough temperature it may turn into liquid phase and all particulars of the metal start to rearrange themselves randomly (van Laarhoven and Aarts 2010). Annealing is the physical process that when cooling the metal from high enough temperature with control all particulars in the metal will rearrange themselves in order to decrease the defects of the metal in the process of minimizing the system energy.

The concept for the method of SA is built on the analogy between this physical annealing process of metal and the problem of solving large combinatorial optimization problems (Chang and Kuo 1994). For the optimization problem of finding the optimal solution with minimum cost, the assumption with equivalences based on the analogy between the physical annealing and optimization problem could be set. In the physical annealing process, the high enough temperature of metal is the initial state of the optimization process. The state of the physical system of the metal at a certain temperature (the arrangement state of all particles of the metal at a certain time point) during the annealing process could be equivalent to a solution of the optimization problem. The temperature reducing process of the metal refers to the slow decrease in the probability of accepting worse solutions in the optimization process. The energy of the physical system of the metal could be equivalent to a cost function of the optimization problem. The lowest energy of the metal refers to the minimum cost which is pursued in the optimization process.

Combined these together, the optimization process is that along with the decrease of the temperature progressively, the algorithm randomly selects a solution close to the current one at each time step, and decides whether to adopt it according to the quality of the solution as well as the temperature-dependent probabilities of selecting better or worse solutions. It allows the selection of worse solutions in the process based on the algorithm for the probability. The detailed explanation of the optimization process with the algorithms is described below, with the illustration as showed in Figure 8.3.

The overall algorithm of SA method could be illustrated with procedures in Figure 8.3. First is to initialize all the parameters related (Figure 8.3, box 1), including starting temperature T_0, decision of total iteration numbers and iteration counter in the beginning $i = 0$. The starting temperature must be high enough for insuring proper searching. A random state is found as the initial state S, which is one of states in the whole search space. The second step is to generate a new state S' randomly from the whole space through a neighborhood structure and a perturbation scheme (Figure 8.3, box 2). Afterwards, it is to calculating and comparing the difference of objective function based on initial state and new state (Figure 8.3, box 3 and box 4). if the new state is better than initial state (objective function is decreased), the new state S' would be accepted and to substitute the initial state as the current state (Figure 8.3, box 5). Otherwise, the Metropolis criterion is applied to judge whether to accept the new state, as showed in Figure 8.3-box 6 with the principle of Eq. 8.5.

$$exp(-\Delta f/T_i) > R; R \in [0, 1] \qquad \text{(Eq. 8.5)}$$

where, $exp(-\Delta f/T_i)$ is the Boltzmann distribution probability; Δf is the difference of objective functions from initial state and new state, as showed in Figure 8.3-box 3; T_i is the current temperature; R represents a randomly generated value from the value range of 0 to 1.

If Eq. 8.5 is satisfied, then accept the new state and go to the next step, otherwise do not accept the new state and go to the next step, which is to judge

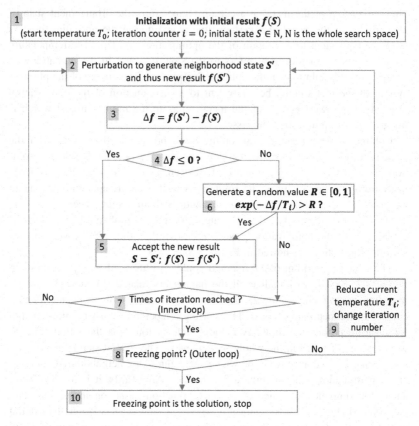

Figure 8.3 Overall flowchart of simulated annealing algorithm (Source: Modified from Chang and Kuo (1994))

whether the condition of inner loop is satisfied. As showed in Figure 8.3-box 7, the condition of inner loop could be that the total iteration number set in the beginning is reached. If the inner loop condition is satisfied, to the further step (Figure 8.3, box 8), otherwise go back to step 2 (Figure 8.3, box 2) to repeat the previous process. The further step is to judge whether the condition of outer loop is satisfied, i.e. whether the temperature is reduced to the freezing point. If yes, go to finial step (Figure 8.3, box 10) and get the solution, otherwise reduce the current temperature by certain principle and also reset the required iteration number (Figure 8.3, box 9) and then go back to step 2.

The key point for simulated annealing is that it designs a principle to accept the worse state than before with a probability, so that the algorithm could jump out of the local optimal trap and search for the global optimal result. This is demonstrated in Figure 8.3 with box 6. The method of simulated annealing is based on random algorithm, which has the ability to get the global optimal result with certain probability but not always. The advantage is that it can get the approximate global optimal result with relatively less time.

Regarding this study, with analogy, the all different possible sets of payment for each measure could be the states in the search space of SA, and the objective function of SA could be the pursuit for the maximum ecological effect with the limitation of given budget. Through the process of SA method, the approximate maximum ecological effect could be achieved within the scope of given budget limitation. At the same time, the corresponding sets of payment for each measure involved in the land use pattern for getting this approximate maximum ecological effect (final solution from SA algorithm) would be present. As a result, the combined information, including the set of payment for each measure, the distribution of measures in heterogeneous HRUs, the given budget and achieved ecological effect, leads to a cost-effective AES design.

8.3   Optimization Modelling Procedure

There are twelve measures and 50 SHUs in this study which are involved in the process of doing the simulation of AES and the optimization of AES. The numbers for both the available measures and the SHUs combined with the complexity of the principles of the simulation and optimization of AES induces that the computer program must be developed for the calculation process. Based on this, the optimization modelling procedure is programed for this study by a computer scientist, Dr. Astrid Sturm, which synthesizes the processes of both the simulation of AES and the optimization of AES described above into a single programmed procedure. The core of the programmed optimization modelling procedure for this study is based on the previously developed program of DSS-Ecopay[5]. DSS-Ecopay is a decision support software tool, designed by Dr. Astrid Sturm and her colleagues, which is programmed for the aim of designing both ecologically effective and cost-effective payments for land use measures in AES to conserve endangered species and habitats in agricultural landscapes (Sturm et al. 2018).

---

[5] For detail, please refer to official website: http://www.inf.fu-berlin.de/DSS-Ecopay/eco pay_main_eng.html.

Although the core is based on DSS-Ecopay, there are many detailed specifications regarding the programming of the optimization modelling procedure for the suitability of the characteristics of this study. First, the environmental aim of this study is the mitigation of soil erosion and water pollution, with the pollutants indices of sediment, nitrogen and phosphorus being targeted, instead of the biodiversity conservation. Second, the area of the SHUs are also different among them in this study. It requires the design of the optimization modelling procedure to be functional in a way that the program could skip a spatial unit which has so big area that makes the budget to be negative and go continuously to check the next spatial unit.

The interface of the programmed optimization modelling procedure in this study is shown in Figure 8.4. With the interface, the available measures in an AES design could be selected among the twelve identified measures in this study; the budget of an AES can be given depending on the needs; and the optimization options refer to the wanted mitigation aim of an AES regarding eight the single pollutant or the various combinations of the different pollutants. In Figure 8.4, it is an example for an AES design with all twelve measures being offered to farmers, the amount of budget being 300,000 RMB, and the mitigation aim being the single pollutant of phosphorus. With the interface, the program is input and operated with all kinds of data from the previous chapters.

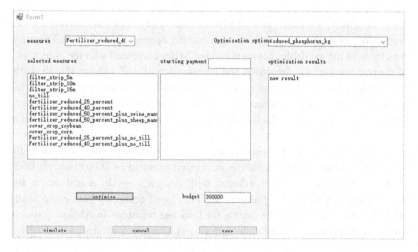

Figure 8.4 Interface of the optimization modelling procedure (Source: Own results)

Results and Analysis 9

With the optimization modelling procedure in Chapter 8, the results of cost-effective AES design could be obtained. This chapter gives the resulted AES of this study based on different scenarios regarding the different requirements for the mitigation targets of AES, budget levels of AES, and involved measures for AES design. For each resulted AES the detailed information is given here, including the budget level, summed payments, and total mitigation effects of the AES, as well as the selected measures, payments and the detailed locations for each of the selected measures.

9.1 Scenarios of AES Design

The optimization modelling procedure is flexible in terms of three perspectives for AES design. With the different combinations for the settings of these three perspectives, there are different scenarios regarding AES design in this study. First, it is flexible for the mitigation targets of an AES based on the three pollutants (sediment, nitrogen, and phosphorus) in this study. Three different kinds of strategic goals for mitigation targets are adopted in this study for AES design. These are mitigation targets regarding each of the individual pollutant (single pollutant of sediment, nitrogen, and phosphorus), and mitigation targets regarding the nutrient pollutants at the same time (the combined targets with nitrogen and phosphorus), as well as mitigation targets regarding all the three pollutants

Supplementary Information The online version contains supplementary material available at https://doi.org/10.1007/978-3-658-41340-8_9.

at the same time (the combined targets with sediment, nitrogen, and phosphorus). Regarding the combination of different pollutants acting as the mitigation targets, different weights are usually given to the different pollutants to show the relative importance of each of the pollutants in the combined mitigation targets. Generally, the weights should be based on the local decision makers and governments' preference (Chen et al. 2015). In the literature involving the best placement of land use measures in watersheds with the combination of pollutant targets, there are research which decide the weights according to the monetary benefits of the unit reduction of each pollutant (e.g. Arabi et al. 2006). Some research give just the simple equal weight for each targeted index (e.g. Chen et al. 2015; Qi and Altinakar 2011). Some other research set different kinds of the weights for each of the pollutants in different scenarios (e.g. Shen et al. 2013).

In this study, the information is unavailable for either the local government's preference for the reduction of each of the three pollutants exactly, or the monetary benefits of unit reduction for each of the three pollutants. The information from the report of the local Environmental Protection Bureau in Xia county is that sediment caused by soil erosion is the most serious problem for the watershed, followed by the nutrient pollutants of nitrogen and phosphorus in sequence (Environmental Protection Bureau in Xia county (EPBX) 2012). Given this, for each of two kinds of the combination targets this study sets three kinds of scenarios regarding the weights for the related pollutants. For the mitigation targets combined with the nutrient pollutants of nitrogen and phosphorus, first scenario is that the equal weights are given to them (50%N+50%P). Second and third scenarios are that more weight is given to nitrogen and less weight for phosphorus but with different weights distribution (60%N+40%P and 70%N+30%P respectively). As to the mitigation targets combined with all the three pollutants, first scenario is also that the equal weight are given to each of them (33%S+33%N+33%P). Second and third scenarios are both hat the maximum weight is given to sediment and less weight to nitrogen and phosphorus in sequence, but with different weight distribution (50%S+30%N+20%P and 60%S+25%N+15%P respectively).

Second, it is flexible for giving the wanted budget limitation for an AES. For each of the mitigation targets this study sets three different kinds of budgets levels for getting the corresponding AES results. The three different budget levels of AES for each mitigation target are set as the ratio of 1:3:5, that are budget level 1 of 100,000 RMB, budget level 2 of 300,000 RMB, and budget level 3 of 500,000 RMB, aiming to observe whether there are regular rules of the resulted land use patterns of the AES design according to the increasing budget level.

Third, it is flexible for giving the involved measures for an AES design based on the identified measures in this study. For each of the mitigation targets, it is decided that there are two sets of involved measures for the AES design in this study. One set of involved measures in AES design refer to all the twelve measures (including both structural measures and operational measures), the other set of involved measures in AES design refer to the nine operational measures in this study, please refer to Table 5.2 in Chapter 5 for the detail of the identified measures. Structural measures in this study refer to the measures of filter strips with different widths. Structural measures are different with operational measures, as these measures are to change the land use of cropland totally to permanent grass. The permanent grass is only removed when the lifetime of AES is ended, which from the perspective of farmers might change the quality of cropland for going back to farming. Therefore, farmers might have different attitude to the structural measures and operational measures when doing the measure selection. Besides, from the perspective of government, at the macro level structural measures induce the reduction of the total area of cropland in the country. However, cropland is rare resource in many regions worldwide, including China especially, for insuring grain supply. Based on these two points, it is meaningful to give two sets of available measures for AES design for each mitigation target, with structural measures and without structural measures.

9.2 Mitigation Targets of Single Pollutant

9.2.1 Sediments

For the mitigation target of single pollutant of sediment, the designed cost-effective AES with twelve measures involved and with nine measures involved, both with three levels of budget limitations, are shown in Figure 9.1 and Figure 9.2 respectively. When there are twelve measures involved, in Figure 9.1, the measures of filter strips (M1, M2, M3) are selected as the major measures under the budget level 1 and budget level 2, while under the budget level 3 the measure of no-till (M4) is mostly selected. When there are nine measures involved in AES design (measures of filter strips, M1, M2, and M3 are excluded), in Figure 9.2, the selected measures are dominated with no-till (M4), especially under the budget level 3. For achieving the cost-effective AES design, the measures which are most cost-effective should be selected mostly in the land use patterns of the resulted AES. According to Figure 6.10 in Chapter 6 for the illustration of the mitigation effects of all identified measures for sediments along with spatial units, in general measures of filters strips and no-till have higher positive mitigation effects compared to all other measures. Meanwhile, according to

Figure 7.8 in Chapter 7 which illustrates the costs situation of all identified measures, the two categories of measures, i.e. filter strips and no-till, have relatively less costs in general along with the SHUs. Combined with these two characteristics, the most cost-effective measures in general for the mitigation effects of sediment are measures of filter strips (M1, M2, and M3) and no-till (M4). Under the situation of the competition of measures of filter strips and no-till, no-till has slightly higher positive mitigation effects along with spatial units (please refer to the data in Appendix 6 in the Electronic Supplementary Material), even when compared with M3 (filter strip of 15 meters) which has best mitigation effects within M1, M2, and M3 due to the longest width of filter strips. Besides, following the principle of the simulation of AES, the spatial units which will be firstly selected by farmers for measures of no-till have better total mitigation effects than that for measures of filter strips[1]. Due to these reasons, the measure of no-till are more frequently selected in the resulted AES than the measures of filter strips, as showed in Figure 9.1 and Figure 9.2.

Along with the increase of the budget level the total mitigation effects of resulted AES increase correspondingly. When budget increases from level 1 to level 2 and level 3, the mitigation percentage of sediment is from 4.2% to 6.8% and 15.9% in Figure 9.1 as well as from 3.4% to 3.5% and 18.1% in Figure 9.2. It is obvious that there is no linear relationship between the amount of budget of AES and its resulted total mitigation effects. The ratio of the total mitigation effects and the summed payments of the land use pattern of AES, from budget level 1 to level 2 and level 3, is from 0.004 to 0.002 and 0.003 in Figure 9.1 as well as from 0.003 to 0.001 and 0.003 in Figure 9.2, with the unit of ton per ha. This shows that there is no rule regarding the cost-effect ratio for the AES results along with the increase of budget level, although bigger budget could lead to higher total mitigation effects. This is due to the principle of the simulation of AES which is from farmers' perspective to firstly select the spatial units for measures with maximum net economic benefit (refer to Eq. 8.1). The first selected spatial units for measures by farmers might not be the ones with good mitigation effects, while during the process of the optimization of AES they must be first selected in order that the ones with good mitigation effects can be then selected later by other farmers for achieving the maximum total mitigation effects.

---

[1] One can see this when adopting the data in Appendix 6 in the Electronic Supplementary Material to do the simulation of AES with the related measures.

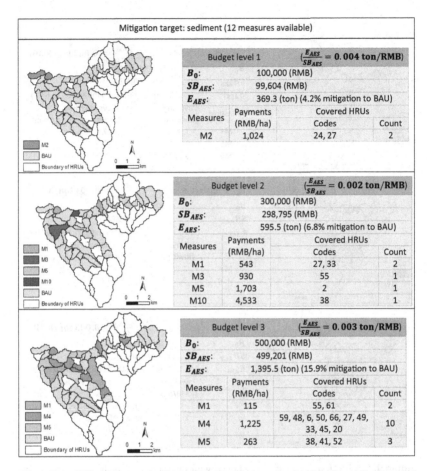

The figure contains three budget-level tables:

Mitigation target: sediment (12 measures available)

Budget level 1	$(\frac{E_{AES}}{SB_{AES}} = 0.004 \text{ ton/RMB})$		
B_0:	100,000 (RMB)		
SB_{AES}:	99,604 (RMB)		
E_{AES}:	369.3 (ton) (4.2% mitigation to BAU)		
Measures	Payments (RMB/ha)	Covered HRUs	
		Codes	Count
M2	1,024	24, 27	2

Budget level 2	$(\frac{E_{AES}}{SB_{AES}} = 0.002 \text{ ton/RMB})$		
B_0:	300,000 (RMB)		
SB_{AES}:	298,795 (RMB)		
E_{AES}:	595.5 (ton) (6.8% mitigation to BAU)		
Measures	Payments (RMB/ha)	Covered HRUs	
		Codes	Count
M1	543	27, 33	2
M3	930	55	1
M5	1,703	2	1
M10	4,533	38	1

Budget level 3	$(\frac{E_{AES}}{SB_{AES}} = 0.003 \text{ ton/RMB})$		
B_0:	500,000 (RMB)		
SB_{AES}:	499,201 (RMB)		
E_{AES}:	1,395.5 (ton) (15.9% mitigation to BAU)		
Measures	Payments (RMB/ha)	Covered HRUs	
		Codes	Count
M1	115	55, 61	2
M4	1,225	59, 48, 6, 50, 66, 27, 49, 33, 45, 20	10
M5	263	38, 41, 52	3

Figure 9.1 AES design results for sediment with three levels of budget (all measures) (Source: Own results, Note: B_0, SB_{AES}, and E_{AES} indicate the given budget, summed payments, and total mitigation effects of the AES (refer to Eq. 8.2, Eq. 8.3, and Eq. 8.4). M1, M2, and M3 are filter strip 5 meters, 10 meters, and 15 meters respectively; M4 is no-till; M5 is chemical fertilizer reduction by 25%; M10 is cover crop of corn (refer to Table 5.2 in Chapter 5). Conversion rate: 1 Euro = 7.7 RMB)

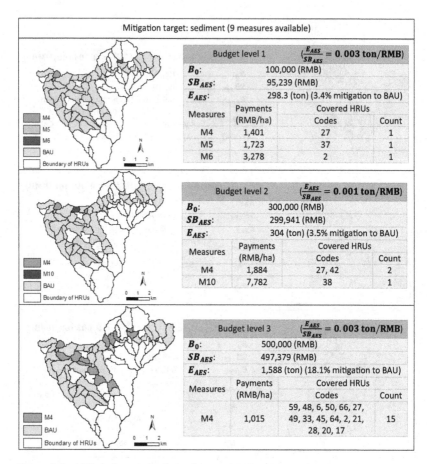

Figure 9.2   AES design results for sediment with three levels of budget (no filter strips) (Source: Own results, Note: B_0, SB_{AES}, and E_{AES} indicate the given budget, summed payments, and total mitigation effects of the AES (refer to Eq. 8.2, Eq. 8.3, and Eq. 8.4). M4 is no-till; M5 and M6 is chemical fertilizer reduction by 25% and 40% respectively; M10 is cover crop of corn (refer to Table 5.2 in Chapter 5). Conversion rate: 1 Euro = 7.7 RMB)

9.2.2 Nitrogen

For the mitigation target of nitrogen, when all measures are involved for AES design, under the budget of level 1, level 2 and level 3 the pollutant load reduced 3.6%, 17.1% and 28.2% respectively compared to BAU, as showed in Figure 9.4. When there are nine measures involved for AES design (without measures of filter strips), the nitrogen load reduced 3.8%, 8.6% and 12.7% under the budget level 1, level 2 and level 3 respectively, as showed in Figure 9.5. It is obvious that when the measures of filter strips (M1, M2, M3) are involved, generally the mitigation effects are much better compared with the situation of that measures of filter strips are missing under the same budget. This is further reflected by the ratios of the total mitigation effects and summed payments of resulted AES, with the ratios of 0.038, 0.061 and 0.62 kg/ha in Figure 9.4 for all twelve measures involved and ratios of 0.651, 0.030 and 0.027 kg/ha in Figure 9.5 for nine measures involved under the budget level 1, level 2 and level 3 respectively[2]. Like the situation with the mitigation target of sediment above, the ratios of total mitigation effects and summed payments of resulted AES for the mitigation target of nitrogen are not changing with obvious rules along with the increase of the budget level. The same reason is due to the principle of the simulation of AES, where farmers select the spatial units for measures based on the net economic benefits they can get from the measures instead of the mitigation effects of measures or the cost-effectiveness of measures.

In Figure 9.4 under each budget level, the mostly selected measures refer to filter strips (M1, M2, M3), especially filter strip of 5 meters (M1). This demonstrates that the measures of filter strips are the most cost-effective ones in general among the twelve identified measures regarding the pollutant reduction of nitrogen. In Figure 9.5, without measures of filter strips, under the three budget levels only measures of no-till (M4) and chemical fertilizer reduction by 25% (M5) are selected, and along with the increase of budget level M5 are selected for the fixed six spatial units under each budget level with the number of spatial units for M4

---

[2] The ratio in Figure 9.5 under the budget level 1, i.e. 0.651 kg/ha, is obvious much bigger than all others. The optimization modelling procedure with the optimization method of simulated annealing can get the approximate global optimal results but might not the optimal results. This high ratio is due to the excellent running results from the optimization modelling procedure, as for this one the summed budget is far away from the budget limitation and with the same scenario setting other tried running results of AES are not as good as this regarding the ratio. However, comparing with other tried AES running results for this scenario, the total mitigation effects do not change too much.

increasing. Therefore, except measures of filter strips, the cost-effective measures for nitrogen would be no-till and chemical fertilizer reduction by 25%.

This could be explained from Figure 6.11 in Chapter 6 for the nitrogen load reduction situation of all measures and from Figure 7.8 in Chapter 7 for the costs of all measures along with SHUs. In Figure 6.11 in Chapter 6, the measures of filter strips (M1, M2, M3), no-till (M4), cover crops (M9, M10), and compounded measures of M11 and M12 have generally better mitigation effects compared to M5, M6, M7, and M8, and their mitigation effects are basically in the similar level. However, in Figure 7.8 in Chapter 7, the abatement costs for M9, M10, M11, and M12 are quite higher than others, and the relatively lower costs in general refer to measures of M1, M2, M3, M4, and M5. Combined with these two sides, it leads to that the measures of filter strips (M1, M2, M3) and no-till (M4) have higher cost-effectiveness and some spatial units for M5. Besides, among these measures, filter strips have better situation for both the mitigation effects and the abatement costs in general along with the spatial units. Among M1, M2 and M3, M1 has the highest ratios of nitrogen mitigation load and abatement costs along with spatial units, as showed in Figure 9.3. This is because M2 and M3 have three times and two times respectively for the abatement costs of M1, while their mitigation effects are not increased as much as their abatement costs. The longer the width of filter strip the lower of the ratio values.

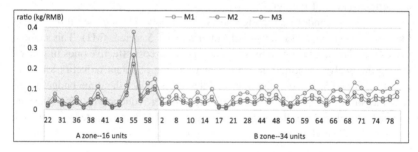

Figure 9.3 Ratio of mitigation load of nitrogen and abatement costs along with spatial units (Source: Own results)

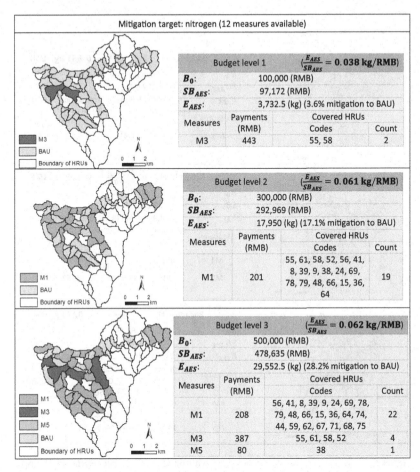

Figure 9.4 AES design results for nitrogen with three levels of budget (all measures) (Source: Own results, Note: B_0, SB_{AES}, and E_{AES} indicate the given budget, summed payments, and total mitigation effects of the AES (refer to Eq. 8.2, Eq. 8.3, and Eq. 8.4). M1 and M3 are filter strip 5 meters and 15 meters respectively; M5 is chemical fertilizer reduction by 25% (refer to Table 5.2 in Chapter 5). Conversion rate: 1 Euro = 7.7 RMB)

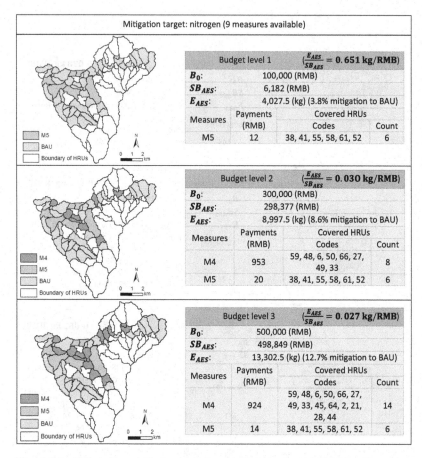

Figure 9.5 AES design results for nitrogen with three levels of budget (no filter strips) (Source: Own results, Note: B_0, SB_{AES}, and E_{AES} indicate the given budget, summed payments, and total mitigation effects of the AES (refer to Eq. 8.2, Eq. 8.3, and Eq. 8.4). M4 is no-till; M5 is chemical fertilizer reduction by 25% (refer to Table 5.2 in Chapter 5). Conversion rate: 1 Euro = 7.7 RMB)

9.2.3 Phosphorus

For the situation with the mitigation target of phosphorus, under the budget level 1, level 2, and level 3, when all twelve measures are involved the resulted AES could reduce the phosphorus load by 4.9%, 22.5%, 41.2% respectively; while when only nine measures are involved the corresponding AES reduce the phosphorus load by 1.8%, 8.2%, and 13.7% respectively. Under the three budget levels, the ratio of total mitigation effects and summed payments of AES are 0.012, 0.018, and 0.020 kg/RMB respectively when all measures involved in Figure 9.7, and are 0.004, 0.007, and 0.006 kg/RMB respectively when measures of filter strips are not involved in Figure 9.8. Like the situation of mitigation target of nitrogen above, without measures of filter strips the cost-effectiveness of AES under the same budget reduced a lot.

As nutrient pollutants, phosphorus and nitrogen have similar existed forms (organic forms and mineral forms) and similar routes for transporting from cropland to waterbodies. Therefore, they have similar situations for the resulted AES regarding the mitigation effects under same budget and the corresponding selected measures, comparing Figure 9.4 and Figure 9.7, as well as Figure 9.5 and Figure 9.8. For phosphorus, as showed in Figure 9.6, when all measures are involved, the measures with best cost-effectiveness refer to filter strips (especially M1) for most of the spatial units and chemical fertilizer reduction by 25% (M5) for several spatial units. Excluding measures of filter strips, the best cost-effective measures refer to M4 and M5 for different spatial units.

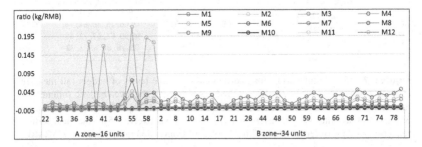

Figure 9.6 Ratio of phosphorus mitigation load and abatement costs along with spatial units (Source: Own results)

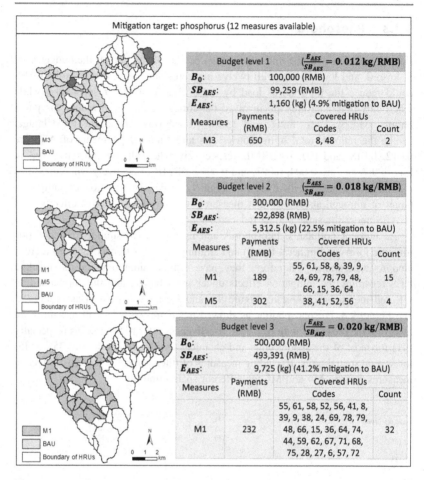

Figure 9.7 AES design results for phosphorus with three levels of budget (all measures) (Source: Own results, Note: B_0, SB_{AES}, and E_{AES} indicate the given budget, summed payments, and total mitigation effects of the AES design (refer to Eq. 8.2, Eq. 8.3, and Eq. 8.4). M1 and M3 are filter strip 5 meters and 15 meters respectively; M5 is chemical fertilizer reduction by 25% (refer to Table 5.2 in Chapter 5). Conversion rate: 1 Euro = 7.7 RMB)

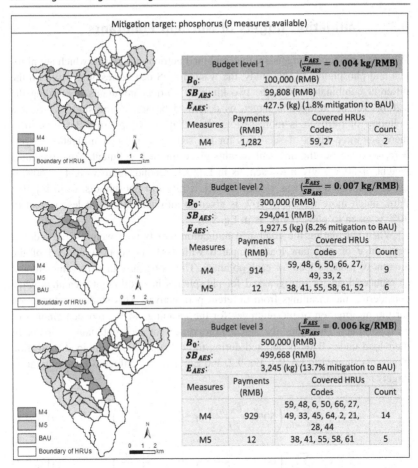

Figure 9.8 AES design results for phosphorus with three levels of budget (no filter strips) (Source: Own results, Note: B_0, SB_{AES}, and E_{AES} indicate the given budget, summed payments, and total mitigation effects of the AES design (refer to Eq. 8.2, Eq. 8.3, and Eq. 8.4). M4 is no-till; M5 is chemical fertilizer reduction by 25% (refer to Table 5.2 in Chapter 5). Conversion rate: 1 Euro = 7.7 RMB)

9.3   Mitigation Targets of Nutrient Pollutants

When the mitigation targets are the combined nutrient pollutants, which are nitrogen and phosphorus in this study, the wanted AES can be resulted based on the different combination with the two kinds of nutrient pollutants. Regarding the settings for the mitigation target, as described before, in the study region the pollution of nitrogen is more serious than that of phosphorus. Based on this, there are three kinds of scenarios for the setting of nutrient mitigation targets here, according to the different weights given for each pollutant. First, the equal weight of 50% is given to each, as in Eq. 9.1; second, more weight of 60% is given to nitrogen and less weight of 40% is given to phosphorus, as in Eq. 9.2; third, much more weight of 70% is given to nitrogen and much less weight of 30% is given to phosphorus, as in Eq. 9.3.

When calculating the combined mitigation targets with weights, the reduced percentage of pollutant load compared with BAU is adopted, instead of the amount of reduced load for each pollutant. This is because that the same amount of reduced load for nitrogen and for phosphorus have different meanings when considering the meanings from different perspectives. This study has limitation for the information of the reference for the meanings of the reduced amount of each pollutant. Therefore, the reduced proportion for each pollutant compared to their load under BAU is adopted during the calculation of combined mitigation targets, as showed in Eq. 9.1, Eq. 9.2, and Eq. 9.3 for the three wanted mitigation targets respectively.

$$E_{AES} = \sum\nolimits_{h_j} \left(50\% \cdot \frac{E_{m_i h_j N}}{L_N} + 50\% \cdot \frac{E_{m_i h_j P}}{L_P} \right) \qquad \text{(Eq. 9.1)}$$

$$E_{AES} = \sum\nolimits_{h_j} \left(60\% \cdot \frac{E_{m_i h_j N}}{L_N} + 40\% \cdot \frac{E_{m_i h_j P}}{L_P} \right) \qquad \text{(Eq. 9.2)}$$

$$E_{AES} = \sum\nolimits_{h_j} \left(70\% \cdot \frac{E_{m_i h_j N}}{L_N} + 30\% \cdot \frac{E_{m_i h_j P}}{L_P} \right) \qquad \text{(Eq. 9.3)}$$

where, E_{AES} refers to the combined mitigation target of a land use pattern in an AES; $E_{m_i h_j N}$ and $E_{m_i h_j P}$ are the amount of reduced pollution load for nitrogen and phosphorus respectively in each spatial unit with the corresponding selected measures, refer to Eq. 8.2 in Chapter 8; L_N and L_P refer to the amount of pollution load under BAU for nitrogen and phosphorus respectively; 50%, 60%, 40%, 70%, and 30% are the weight factors respectively, representing the relative importance of the reduction of each pollutant in the AES design; h_j refers to the SHUs; m_i refers to the measures selected for the corresponding spatial units; i and j refer to the code of different measures and the code of SHUs respectively.

9.3.1 Twelve Measures

When twelve measures are involved, the resulted AES design for the three kinds of combined mitigation targets with three levels of budget are shown in Figure 9.10, Figure 9.11 and Figure 9.12 respectively. According to these figures, it is clear that under each budget level, the resulted mitigation effects (pollution load reduction percentage compared to BAU) are more or less the same between the three combined mitigation targets. Beside, these resulted AES with combined mitigation targets are similar like the ones under the same budget for the mitigation targets of nitrogen and phosphorus (refer to Figure 9.4 and Figure 9.7), in terms of both the mitigation effects and the selected measures.

This is because the most cost-effective measures for either nitrogen or phosphorus are the same ones, which are M1 (filter strips 5 meters) for most of the spatial units and M5 (chemical fertilizer reduction by 25%) for several spatial units. The selected measures for both the combined nutrient mitigation targets (refer to Figure 9.10, Figure 9.11 and Figure 9.12) and the single mitigation targets of nitrogen and phosphorus (refer to Figure 9.4 and Figure 9.7) are mostly M1 and then slightly M5, M2, and M3.

Besides, for each of these measures (M1, M2, M3, and M5), the ratios of reduced pollution load and abatement costs regarding nitrogen and phosphorus have the same changing trends along the SHUs, as illustrated in Figure 9.9. This demonstrates that, for each of these measures, in a certain spatial unit when the measure has higher (or lower) mitigation effects for nitrogen it also has higher (or lower) mitigation effects for phosphorus. Therefore, although the three kinds of combined nutrient mitigation targets have different weights for nitrogen and phosphorus (Figure 9.10, Figure 9.11 and Figure 9.12), the different weight distributions actually cannot influence the mitigation effects of each pollutant from the resulted AES much. As when the optimization modelling procedure selects the most cost-effective measures and corresponding spatial units for nitrogen mitigation, the selected ones are also the most cost-effective measures and spatial units for the mitigation of phosphorus, as illustrated in Figure 9.9.

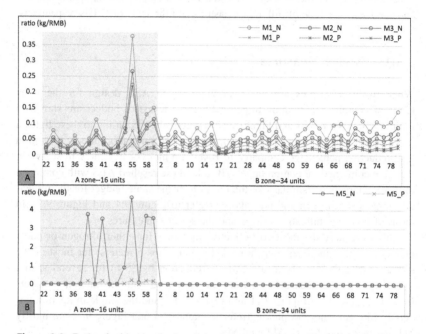

Figure 9.9 Ratio of mitigation load and abatement costs along with spatial units (Source: Own results, Note: M1_N, M2_N, M3_N, and M5_N refer to the situations of corresponding measures for nitrogen; M1_P, M2_P, M3_P, and M5_P refer to the situations of corresponding measures for phosphorus)

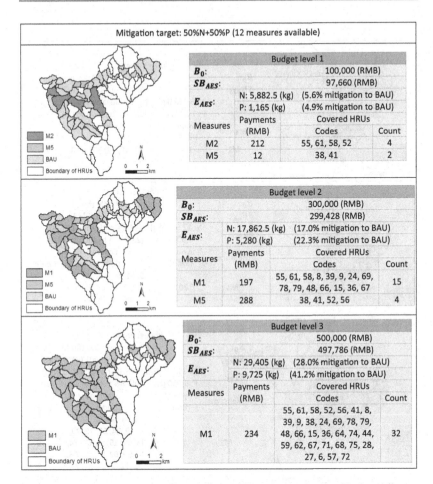

Figure 9.10 AES design results for "50%N+50%P" with three levels of budget (all measures) (Source: Own results, Note: B_0, SB_{AES}, and E_{AES} indicate the given budget, summed payments, and total mitigation effects of the AES design (refer to Eq. 8.2, Eq. 8.3, and Eq. 8.4). M1 and M2 are filter strip 5 meters and 10 meters respectively; M5 is chemical fertilizer reduction by 25% (refer to Table 5.2 in Chapter 5). Conversion rate: 1 Euro = 7.7 RMB)

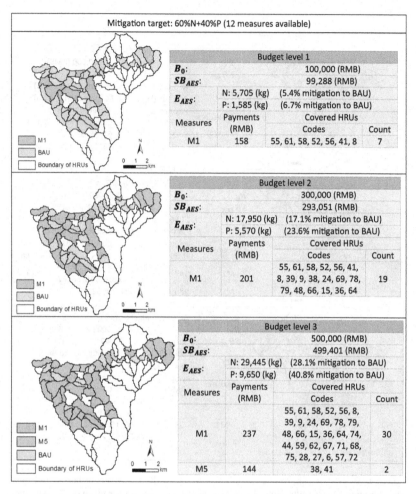

Figure 9.11 AES design results for "60%N+40%P" with three levels of budget (all measures) (Source: Own results, Note: B_0, SB_{AES}, and E_{AES} indicate the given budget, summed payments, and total mitigation effects of the AES design (Eq. 8.2, Eq. 8.3, and Eq. 8.4). M1 is filter strip 5 meters; M5 is chemical fertilizer reduction by 25% (refer to Table 5.2 in Chapter 5). Conversion rate: 1 Euro = 7.7 RMB)

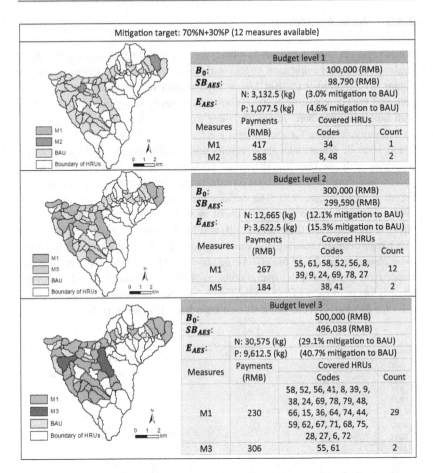

Mitigation target: 70%N+30%P (12 measures available)			
Budget level 1			
B_0:		100,000 (RMB)	
SB_{AES}:		98,790 (RMB)	
E_{AES}:	N: 3,132.5 (kg)	(3.0% mitigation to BAU)	
	P: 1,077.5 (kg)	(4.6% mitigation to BAU)	
Measures	Payments (RMB)	Covered HRUs	
		Codes	Count
M1	417	34	1
M2	588	8, 48	2
Budget level 2			
B_0:		300,000 (RMB)	
SB_{AES}:		299,590 (RMB)	
E_{AES}:	N: 12,665 (kg)	(12.1% mitigation to BAU)	
	P: 3,622.5 (kg)	(15.3% mitigation to BAU)	
Measures	Payments (RMB)	Covered HRUs	
		Codes	Count
M1	267	55, 61, 58, 52, 56, 8, 39, 9, 24, 69, 78, 27	12
M5	184	38, 41	2
Budget level 3			
B_0:		500,000 (RMB)	
SB_{AES}:		496,038 (RMB)	
E_{AES}:	N: 30,575 (kg)	(29.1% mitigation to BAU)	
	P: 9,612.5 (kg)	(40.7% mitigation to BAU)	
Measures	Payments (RMB)	Covered HRUs	
		Codes	Count
M1	230	58, 52, 56, 41, 8, 39, 9, 38, 24, 69, 78, 79, 48, 66, 15, 36, 64, 74, 44, 59, 62, 67, 71, 68, 75, 28, 27, 6, 72	29
M3	306	55, 61	2

Figure 9.12 AES design results for "70%N+30%P" with three levels of budget (all measures) (Source: Own results, Note: B_0, SB_{AES}, and E_{AES} indicate the given budget, summed payments, and total mitigation effects of the AES design (refer to Eq. 8.2, Eq. 8.3, and Eq. 8.4). M1, M2, and M3 are filter strips with different widths; M5 is chemical fertilizer reduction by 25% (refer to Table 5.2 in Chapter 5). Conversion rate: 1 Euro = 7.7 RMB)

9.3.2   Nine Measures

When there are nine measures involved, the resulted AES design for the three
kinds of combined mitigation targets with three levels of budget are shown in
Figure 9.14, Figure 9.15, and Figure 9.16 respectively. Similar like the situation with twelve measures involved above, the resulted mitigation effects of AES
with nine measures here among the three kinds of combined mitigation targets
are basically the same for each budget level, especially for higher budget of level
2 and level 3. Meanwhile, these results under each budget level have similar
situations with the ones for the single pollutant targets of nitrogen and phosphorus regarding both the mitigation effects and the selected measures (refer to
Figure 9.5 and Figure 9.8). The reasons are the same like the situations with
twelve measures involved above. Without measures of filter strips, the most cost-
effective measures are then M4 (no-till) and M5 (chemical fertilizer reduction by
25%) regarding different spatial units respectively. Both M4 and M5 have the
same changing trends for the effect-cost ratio regarding nitrogen and phosphorus
respectively along with the spatial units, as showed in Figure 9.13. M5 has very
low payments in each of the resulted AES, due to that there are several spatial
units for M5 have very high effect-cost ratio but others are very low.

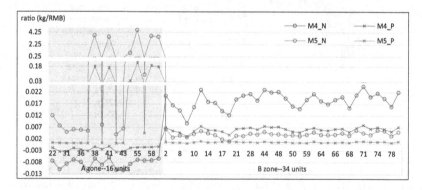

Figure 9.13 Ratio of mitigation load and abatement costs along with spatial units (Source:
Own results, Note: M4_N and M5_N refer to the situations of corresponding measures for
nitrogen; M4_P and M5_P refer to the situations of corresponding measures for phosphorus)

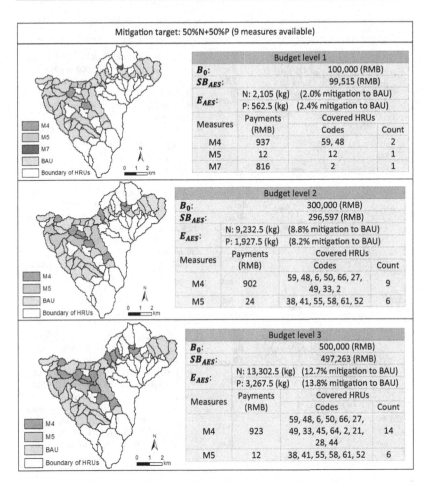

Figure 9.14 AES design results for "50%N+50%P" with three levels of budget (no filter strips) (Source: Own results, Note: B_0, SB_{AES}, and E_{AES} indicate the given budget, summed payments, and total mitigation effects of the AES design (refer to Eq. 8.2, Eq. 8.3, and Eq. 8.4). M4 is no-till; M5 is chemical fertilizer reduction by 25%; M7 is chemical fertilizer reduction by 50% along with 1000kg/ha swine manure application (refer to Table 5.2 in Chapter 5). Conversion rate: 1 Euro = 7.7 RMB)

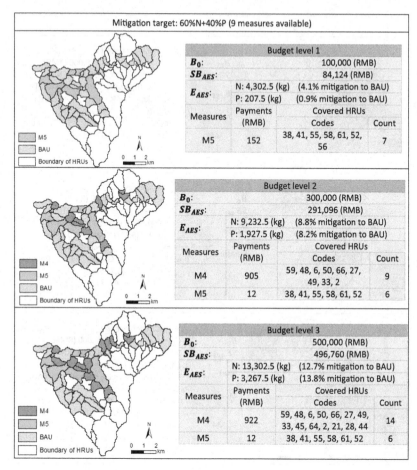

Figure 9.15　AES design results for "60%N+40%P" with three levels of budget (no filter strips) (Source: Own results, Note: B_0, SB_{AES}, and E_{AES} indicate the given budget, summed payments, and total mitigation effects of the AES design (refer to Eq. 8.2, Eq. 8.3, and Eq. 8.4). M4 is no-till; M5 is chemical fertilizer reduction by 25% (refer to Table 5.2 in Chapter 5). Conversion rate: 1 Euro = 7.7 RMB)

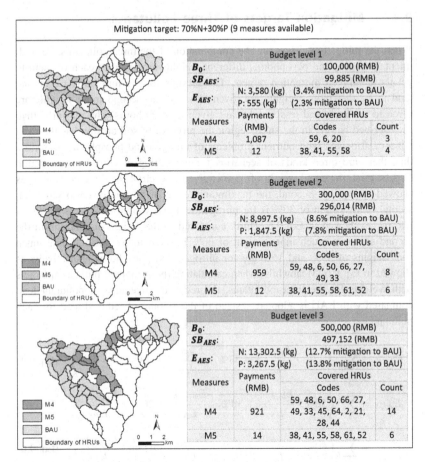

Figure 9.16 AES design results for "70%N+30%P" with three levels of budget (no filter strips) (Source: Own results, Note: B_0, SB_{AES}, and E_{AES} indicate the given budget, summed payments, and total mitigation effects of the AES design (refer to Eq. 8.2, Eq. 8.3, and Eq. 8.4). M4 is no-till; M5 is chemical fertilizer reduction by 25% (refer to Table 5.2 in Chapter 5). Conversion rate: 1 Euro = 7.7 RMB)

9.4   Mitigation Targets of Three Pollutants

When the mitigation targets are the combination of the three pollutants, i.e. sediment, nitrogen, and phosphorus, different weights are given to each of them to get the combined mitigation targets. The settings for the mitigation targets are tend to be that more weights are given to sediment, second more weights are given to nitrogen, and less weights are given to phosphorus. The reason is that, as described before, in the study region the problem of soil erosion has the most concern, followed with nutrient problems of nitrogen and phosphorus. Based on this, there are three kinds of scenarios for the setting of combined mitigation targets. First, the equal weight of 33% is given to each of the three pollutants, as showed in Eq. 9.4. Second, the maximum weight of 50% is given to sediment, second more weight of 30% is for nitrogen and less weight of 20% is for phosphorus, as showed in Eq. 9.5. Third, compared to the weight distribution in the second scenario, much more weight of 60% is given for sediment, while much less weights of 25% and 15% are for nitrogen and phosphorus respectively, as showed in Eq. 9.6. Meanwhile, when calculating the combined mitigation targets with weights for the three pollutants, the reduced percentage of pollutant load compared with BAU is adopted, instead of the amount of reduced load for each pollutant.

$$E_{AES} = \sum_{h_j} \left(33\% \cdot \frac{E_{m_i h_j S}}{L_S} + 33\% \cdot \frac{E_{m_i h_j N}}{L_N} + 33\% \cdot \frac{E_{m_i h_j P}}{L_P} \right) \quad \text{(Eq. 9.4)}$$

$$E_{AES} = \sum_{h} \left(50\% \cdot \frac{E_{m_i h_j S}}{L_S} + 30\% \cdot \frac{E_{m_i h_j N}}{L_N} + 20\% \cdot \frac{E_{m_i h_j P}}{L_P} \right) \quad \text{(Eq. 9.5)}$$

$$E_{AES} = \sum_{h} \left(60\% \cdot \frac{E_{m_i h_j S}}{L_S} + 25\% \cdot \frac{E_{m_i h_j N}}{L_N} + 15\% \cdot \frac{E_{m_i h_j P}}{L_P} \right) \quad \text{(Eq. 9.6)}$$

where, E_{AES} refers to the combined mitigation target for AES design; $E_{m_i h_j S}$, $E_{m_i h_j N}$ and $E_{m_i h_j P}$ are the amount of reduced pollution load for sediment, nitrogen and phosphorus respectively in each spatial unit with the corresponding selected measures (refer to Eq. 8.2 in Chapter 8); L_S, L_N and L_P refer to the amount of pollution load under BAU for sediment, nitrogen and phosphorus respectively; h_j refers to the SHUs; m_i refers to the measures selected for the corresponding spatial units; i and j refer to the code of different measures and the code of SHUs respectively.

9.4.1 Twelve Measures

The resulted AES for the three settings of combination targets with all measure involved, under three budget levels, are presented in Figure 9.18, Figure 9.19 and Figure 9.20 respectively. It is observed that under the budget level 3, the resulted AES for the three settings of combination targets are the same, except the amount of payment for measures and then summed payments. Under the budget level 2, along with the increase for the weight of sediment and decrease for the weights of nitrogen and phosphorus, the mitigation effects for both nitrogen and phosphorus are reduced, while the mitigation effects for sediments is not consistently increased. Under the budget level 1, along with the increase for the weight of sediment and decrease for the weights of nitrogen and phosphorus, the mitigation effects for sediments increased consistently.

According to Figure 9.18, Figure 9.19 and Figure 9.20, the majority of spatial units are selected with M1 (filer strip of 5 meters). This is because for either sediment, nitrogen, or phosphorus, filter strips (M1, M2, M3) are the generally most cost-effective measures. Among M1, M2, and M3, M1 has the highest ratio of mitigation effects and costs for either sediment (Figure 9.17), or nitrogen and phosphorus (Figure 9.9). Besides, filter strips are the only measures in this study which do not have negative values for either sediment, nitrogen, or phosphorus.

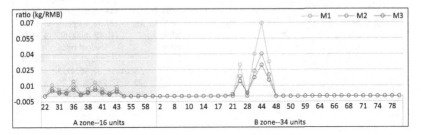

Figure 9.17 Ratio of sediment mitigation load and abatement costs along with spatial units (Source: Own results)

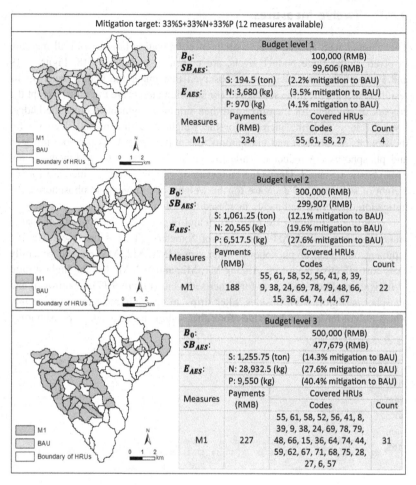

Mitigation target: 33%S+33%N+33%P (12 measures available)			
Budget level 1			
B_0:	100,000 (RMB)		
SB_{AES}:	99,606 (RMB)		
E_{AES}:	S: 194.5 (ton)	(2.2% mitigation to BAU)	
	N: 3,680 (kg)	(3.5% mitigation to BAU)	
	P: 970 (kg)	(4.1% mitigation to BAU)	
Measures	Payments (RMB)	Covered HRUs	
		Codes	Count
M1	234	55, 61, 58, 27	4
Budget level 2			
B_0:	300,000 (RMB)		
SB_{AES}:	299,907 (RMB)		
E_{AES}:	S: 1,061.25 (ton)	(12.1% mitigation to BAU)	
	N: 20,565 (kg)	(19.6% mitigation to BAU)	
	P: 6,517.5 (kg)	(27.6% mitigation to BAU)	
Measures	Payments (RMB)	Covered HRUs	
		Codes	Count
M1	188	55, 61, 58, 52, 56, 41, 8, 39, 9, 38, 24, 69, 78, 79, 48, 66, 15, 36, 64, 74, 44, 67	22
Budget level 3			
B_0:	500,000 (RMB)		
SB_{AES}:	477,679 (RMB)		
E_{AES}:	S: 1,255.75 (ton)	(14.3% mitigation to BAU)	
	N: 28,932.5 (kg)	(27.6% mitigation to BAU)	
	P: 9,550 (kg)	(40.4% mitigation to BAU)	
Measures	Payments (RMB)	Covered HRUs	
		Codes	Count
M1	227	55, 61, 58, 52, 56, 41, 8, 39, 9, 38, 24, 69, 78, 79, 48, 66, 15, 36, 64, 74, 44, 59, 62, 67, 71, 68, 75, 28, 27, 6, 57	31

Figure 9.18 AES with the mitigation target of "33%S+33%N+33%P" (all measures) (Source: Own results, Note: B_0, SB_{AES}, and E_{AES} indicate the given budget, summed payments, and total mitigation effects of the AES design (refer to Eq. 8.2, Eq. 8.3, and Eq. 8.4). M1 is filter strip of 5 meters (refer to Table 5.2 in Chapter 5). Conversion rate: 1 Euro = 7.7 RMB)

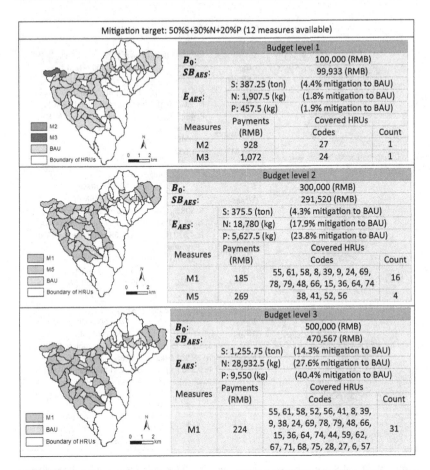

Figure 9.19 AES with the mitigation target of "50%S+30%N+20%P" (all measures) (Source: Own results, Note: B_0, SB_{AES}, and E_{AES} indicate the given budget, summed payments, and total mitigation effects of the AES design (refer to Eq. 8.2, Eq. 8.3, and Eq. 8.4). M1, M2, and M3 are filter strips with different widths; M5 is chemical fertilizer reduction by 25% (refer to Table 5.2 in Chapter 5). Conversion rate: 1 Euro = 7.7 RMB)

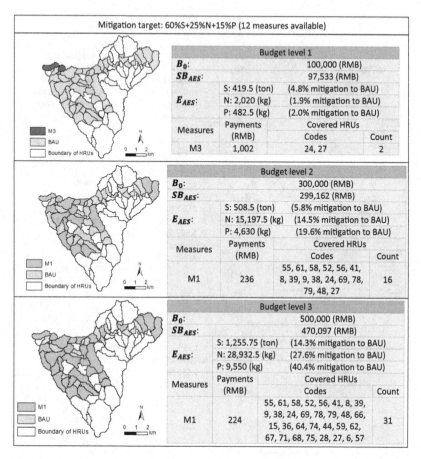

Figure 9.20 AES with the mitigation target of "60%S+25%N+15%P" (all measures) (Source: Own results, Note: B_0, SB_{AES}, and E_{AES} indicate the given budget, summed payments, and total mitigation effects of the AES design (refer to Eq. 8.2, Eq. 8.3, and Eq. 8.4). M1 and M3 are filter strips with different widths; M5 is chemical fertilizer reduction by 25% (refer to Table 5.2 in Chapter 5). Conversion rate: 1 Euro = 7.7 RMB)

9.4.2 Nine Measures

When there are only nine measures involved for the AES design, the results are shown in Figure 9.21, Figure 9.22, and Figure 9.23 for the three kinds of combination targets respectively with each has three budgets levels. According to these results, although with different weights for the three pollutants of the combined mitigation targets, under the budget level 1 and level 3 the resulted AES with mitigation targets of "33%S+33%N+33%P" and "50%S+30%N+20%P" have the same land use patterns in terms of the selected measures and the corresponding spatial units for the measures (payments for the measures and then the summed payments of the AES are different). With mitigation target of "60%S+25%N+15%P" with increased weight for sediment and decreased weights for nitrogen and phosphorus compared to the other two mitigation targets here, for the AES under both budget level 1 and budget level 3 the reduced load for sediment increased and the reduced load for nitrogen and phosphorus decreased. Under budget level 2, for the AES with the three kinds of mitigation targets here, the reduced load of sediment keeps the same and the reduced load for nitrogen and phosphorus changed more or less.

It is obvious that when there are nine measures involved for the AES design with the three kinds of combined mitigation targets here, only measures of M4 (no-till) and M5 (chemical fertilizer reduction by 25%) are selected (refer to Figure 9.21, Figure 9.22, and Figure 9.23). This is because that, as described before, without measures filter strips, the most cost-effective measures for sediment are then M4 (refer to the situation of AES results for the single mitigation target of sediment). Meanwhile, without measures of filter strips, the most cost-effective measures for both nitrogen and phosphorus are then M4 and M5 (refer to the situation of AES results for the single mitigation target of nitrogen and phosphorus). Besides, from the resulted AES in Figure 9.21, Figure 9.22, and Figure 9.23, it is noticed that the corresponding spatial units for M5 and M4 have their relatively fixed locations respectively under different budget levels and with different mitigation targets. This is because that M4 and M5 have their different corresponding spatial units respectively for being cost-effective regarding the mitigation of nitrogen and phosphorus (refer to Figure 9.13).

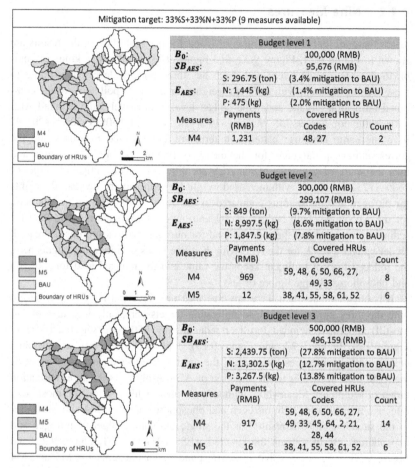

Figure 9.21 AES with the mitigation target of "33%S+33%N+33%P" (no filter strips) (Source: Own results, Note: B_0, SB_{AES}, and E_{AES} indicate the given budget, summed payments, and total mitigation effects of the AES design (refer to Eq. 8.2, Eq. 8.3, and Eq. 8.4). M4 is no-till; M5 is chemical fertilizer reduction by 25% (refer to Table 5.2 in Chapter 5). Conversion rate: 1 Euro = 7.7 RMB)

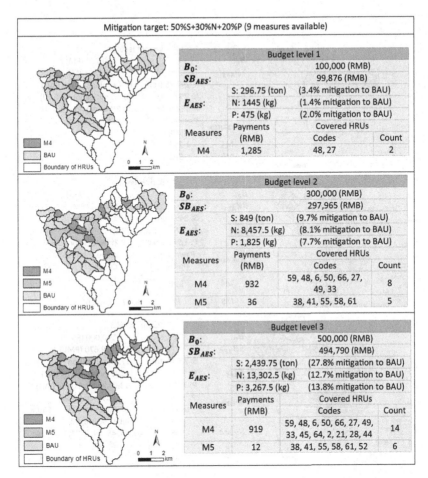

Figure 9.22 AES with the mitigation target of "50%S+30%N+20%P" (no filter strips) (Source: Own results, Note: B_0, SB_{AES}, and E_{AES} indicate the given budget, summed payments, and total mitigation effects of the AES design (refer to Eq. 8.2, Eq. 8.3, and Eq. 8.4). M4 is no-till; M5 is chemical fertilizer reduction by 25% (refer to Table 5.2 in Chapter 5). Conversion rate: 1 Euro = 7.7 RMB)

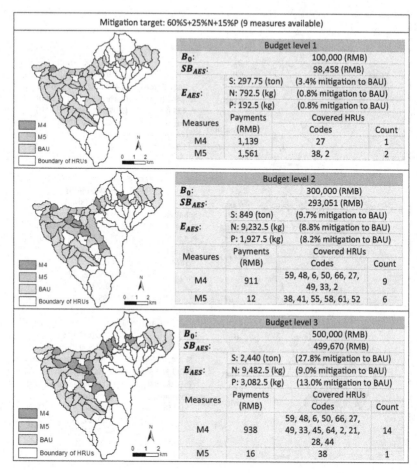

Figure 9.23 AES with the mitigation target of "60%S+25%N+15%P" (no filter strips) (Source: Own results, Note: B_0, SB_{AES}, and E_{AES} indicate the given budget, summed payments, and total mitigation effects of the AES design (refer to Eq. 8.2, Eq. 8.3, and Eq. 8.4). M4 is no-till; M5 is chemical fertilizer reduction by 25% (refer to Table 5.2 in Chapter 5). Conversion rate: 1 Euro = 7.7 RMB)

9.5   Relationship of Budget Size and Pollution Mitigation

Regarding the relationship between budget size and resulting pollution mitigation, one would expect that budget increases also lead to more mitigation, but with mitigation increases decreasing with rising budget levels (positive but decreasing marginal pollution reduction benefits). The logic for this is that with increasing budget size, it is expected that there are more spatial units being selected which bring pollution mitigation effects along with them. It is also expected, however, that the more cost-effective spatial units and their attached measures are selected first with less cost-effective spatial units and their attached measures being selected later during the simulation and optimization process. Based on these two points of understanding, it would be that along with increasing budget size, there is increasing pollution mitigation effects but the marginal increasing effects of the pollution mitigation is decreasing.

However, this expectation might not always be consistent with the situation resulted. The relationship between budget size and the resulted pollution mitigation effects of AES are shown in Figure 9.24, with the results of some representative AES from above sections regarding the situations for each of the pollutants (S, N, and P). If as expected, the marginal increasing effect of pollution mitigation is decreasing along with the rise of budget size, then there should be concave increasing curves for mitigation effects. However, in Figure 9.24 the shapes of the resulted curves are irregular (not concave) for either pollutant of S, N, or P. In many situations, when there are increasing budgets, the marginal mitigation effects of pollutant are increasing instead of decreasing. For example, comparing the pollution mitigation effect changes with a budget rise from 100,000 RMB to 300,000 RMB and a budget rise from zero to 100,000 RMB, all scenarios in Figure 9.24-B and in Figure 9.24-C have increasing marginal mitigation effects with increasing budget size. The same situation happens in Figure 9.24-A for scenarios of S_12, S_9, and 33%S+33%N+33%P_9, comparing a budget rise from 300,000 RMB to 500,000 RMB and a budget rise from 100,000 RMB to 300,000 RMB.

The irregular relationships between budget size and environmental effects also appeared in previous research, such as Ando et al. (1998). For this study, two reasons might explain the increasing marginal pollution mitigation effects, which are the different sizes of the HRUs and the principle of spatial unit selection of AES simulation considering only cost. The explanation is based on the integrated effects of these two reasons.

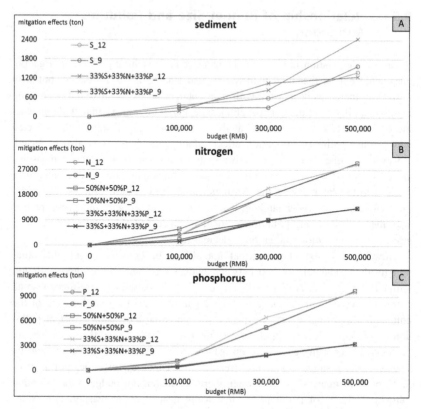

Figure 9.24 Relationship between budget size and resulting pollution mitigation (Source: Own results, Note: S_12, N_12, and P_12 refer to scenarios for mitigation targets of single pollutant of S, N, and P with 12 measures (corresponding to Figure 9.1, Figure 9.4, and Figure 9.7 respectively). S_9, N_9, and P_9 refer to scenarios for mitigation targets of single pollutant of S, N, and P with 9 measures (corresponding to Figure 9.2, Figure 9.5, and Figure 9.8 respectively). 50%N+50%P_12 and 50%N+50%P_9 refer to scenarios for mitigation targets of combined nutrient pollutants with 12 measures and 9 measures respectively (corresponding to Figure 9.10 and Figure 9.14 respectively). 33%S+33%N+33%P_12 and 33%S+33%N+33%P_9 refer to scenarios for mitigation targets of combined all three pollutants with 12 measures and 9 measures respectively (corresponding to Figure 9.18 and Figure 9.21 respectively))

During the simulation and optimization process of AES, although the most cost-effective measures are first considered, the SHUs firstly selected are not necessarily the SHUs that have higher cost-effectiveness of this measure. As during the process of AES simulation, the SHUs with the least unit area costs are first considered for selection. However, a SHU with very low unit area cost might have a very large size of area, which might result in a high total cost and then payment for this SHU. Meanwhile, it is possible that a SHU with a low unit area cost may have also low mitigation effects for pollutants. If this is the case, then the first selected SHUs (with low unit area costs) are the SHUs with low cost-effectiveness (as having low mitigation effects).

This triggers a possible situation that under a low budget all the selected SHUs along with the attached measures are generally not cost-effective. Because that due to the simulation principle the SHUs with higher mitigation impacts have too high unit area costs to be ranked for the selection possibility under this low budget. Or because that the SHUs with higher mitigation impacts can be ranked for selection under the budget level, but these SHUs have too big size of area to make the total payment to be less than the budget. Therefore, these (more cost-effective) SHUs are skipped under the low budget level, and can only be selected under a higher budget level. Both of the situations induce the result of non-decreasing marginal pollution mitigation effects along with the increase of budget size.

Conclusion and Discussion

10

This chapter aims to make a summary for this study, along with the discussion for the limitations and possible future research of this study, as well as the policy recommendations for the government based on this study. Summary of this study focuses on the highlights and major innovations of this study. Part of limitations and future research describes the major limitations of this study and the possible future research directions based on these limitations. Policy recommendations gives the recommendations for the government regarding the important points for cost-effective AES design in general and the specified suggestions for the local watershed environmental protection of the study region.

10.1 Summary

(I) Developing a novel generic method

It is not new in the research to combine ecological, economic components and optimization process to solve water pollution problems in a watershed in a cost-effective way. Some applies relatively simply ways to deal with optimization process, such as ranking the measures in terms of the ratio of ecological effect and economic cost of the measures (e.g. Lescot et al. 2013; Turpin et al. 2005; Cools et al. 2011). Some adopts complex heuristic search algorithms to go through the optimization process to get the optimum combination of measure allocations (e.g. Qi and Altinakar 2011; Arabi et al. 2006; Maringanti et al. 2011; Rodriguez et al.

Supplementary Information The online version contains supplementary material available at https://doi.org/10.1007/978-3-658-41340-8_10.

Z. Hao, *An Integrated Modelling Approach to Design Cost-Effective AES for Agricultural Soil Erosion and Water Pollution*, https://doi.org/10.1007/978-3-658-41340-8_10

2011). Besides, there are research which concludes the general key components and sequential steps for solving the problem, based on the integrated ecological-economic procedures for the analysis of cost-effectiveness of agri-environmental measures for water quality in EU (Balana et al. 2015; Balana et al. 2011). However, these studies are based on a planning perspective, which do not consider the cooperation of farmers for the application of the study results in practice. Meanwhile, these studies do not consider the spatial heterogeneity of economic costs of measures the same as the consideration for the spatial heterogeneity of ecological effects of measures. These are the two points of major gaps in the previous research, as explained in detail below.

a. Although the spatial heterogeneity is emphasized and analyzed in terms of ecological effects in most of the studies, the spatial heterogeneity in the perspective of economic cost is generally ignored. However, in reality the economic costs of measures in different spatial units are usually distinguished due to lots of factors. For pursuing the cost-effectiveness regarding the distribution of land use measures, the information of heterogeneous costs should be obtained and operated during the optimization process at the same level in terms of the SHUs as that for the analysis of the ecological effects of measures.

b. In practice, the one who would finally implement the agri-environmental measures are land users, which are farmers in most of the situation. To get the better cooperation of farmers for implementing the measures, farmers should have the voluntary right to decide whether to implement measures, which measures to implement and in where. This requirement is one of the key features of AES. Therefore, to design the cost-effective land use patterns with the distribution of measures under the AES principles is very important and meaningful.

This study contributed to filling these gaps. The aim of the study was to develop a novel generic method, with which the cost-effective AES can be designed according to the expectations for the AES regarding the budget limitations and environmental targets. It developed an integrated hydro-economic modelling procedure, which had components of agri-environmental measures selection, eco-hydrological model simulation, economic cost assessment, and the simulation and optimization of AES. Each of these components had its own independent tasks but these tasks must be completed with the coordination and cooperation with each of the other components. Under this integrated modelling procedure, the spatial heterogeneity in terms of both ecological effects and economic costs of measures are considered and assessed in the same level. The intermediate results of the heterogeneous information for the ecological effects and economic costs of measures are combined for doing the simulation and optimization of AES in order to get the cost-effective AES design. In this process, the simulation for famer's selection behavior with different measures is completely considered and combined into the optimization process for getting the optimum land use patterns in terms of cost-effectiveness.

The developed systematic method with the ecological-economic modelling procedure was able to identify the cost-effective AES regarding the concerned NPS water pollution mitigation. The power of the method is demonstrated in this study by its application in a real watershed in Shanxi province in China. However, the developed method is generic, as it can be applied flexibly to all kinds of agricultural watersheds with different characteristics and areas. Meanwhile, it is feasible to consider other NPS pollutants (other than sediment, nitrogen, and phosphorus in this study), like pesticides and other related chemicals with the corresponding land use measures (as long as the pollutants could be simulated by the corresponding eco-hydrological model which is adopted in the hydro-economic modelling procedure of the developed method).

(II) Interdisciplinary research

Interdisciplinary research was the feature and also the challenge of this study. The developed integrated hydro-economic modelling procedure involved multi-disciplines, including ecology and hydrology (eco-hydrological modelling), economics (opportunity costs assessment), agronomy (agri-environmental measure identification), and mathematical optimization (optimization modelling procedure). These different disciplines were applied for carrying out the tasks of each of the components in the integrated modelling procedure, as described in below.

a. Regarding identifying the proper agri-environmental measures which are suitable to the study region and for the targeted pollutants, agronomic knowledge is involved (Cai et al. 2003; Brouwer and Hofkes 2008). As the measures should be implemented on cropland along with the crop production, which makes that the familiarity of detailed procedures for crop production from sowing to harvest for each related crop type is needed. Meanwhile, these measures were needed to input into SWAT model for simulating their induced changes of crop yield and changes of pollution load. This involved the knowledge of relationships between influenced factors for crop yield and the generation of pollution load for each pollutant during the cropping process.

b. In this study the eco-hydrological model was needed for simulating the quantified effects of each measure in each heterogeneous spatial unit, which made the ecological and hydrological knowledge to be involved. This is because that for simulating the pollution load caused by water, the eco-hydrological model needed to combine the knowledge of hydrological principles and ecological principles regarding NPS water pollution. As an eco-hydrological model, SWAT was adopted in this study. The design of SWAT model itself combined lots of interactions between hydrological and ecological systems, such as nutrient cycling along with hydrological cycling (Essenfelder et al. 2018). To build up SWAT model for the simulation aim, the hydrological knowledge and involved relationships with different ecological results of NPS pollutants are required.

c. For assessing the costs of each measure in each SHUs, economics was there. First, the general cost categories of AES from farmers' perspective were analyzed. Second, focused on the abatement cost of each measure, the subcategories of abatement cost were analyzed and were identified for each measure. Third, based on the changed crop production procedures caused by each measure, the detailed cost calculation formulas are established respectively for each of the measures. Besides, during the calculation process for the abatement cost of each measure, knowledge of discounting over multiple years for obtaining the average annual cost of each measure in each spatial unit, as well as questionnaire design, survey and statistical analysis for the data are involved.

d. Regarding the step of "simulation and optimization of AES" in the integrated modelling procedure, knowledge of numerical optimization process as well as related mathematical optimization were demanded. The principles for the simulation of AES was analyzed and the optimization method of simulated annealing was adopted for the optimization of AES in this study. Both the calculations for simulation of AES and the optimization of AES were very complex, which must be processed with the help of mathematical optimization. The processes for the simulation of AES and for the optimization of AES were integrated into a single optimization modelling procedure, which were developed by a computer expert.

With interdisciplinary research, the challenge of this study was to make the interdisciplinary components in the integrated modelling procedure to be coordinated and cooperated in order to achieve the final research aim. It is a general difficulty for all relevant studies coupling ecology and economics (Cools et al. 2011). In this study, to make the coordination and cooperation of different components in different disciplines, first the analysis for the interrelations between each of the two components were performed, in order to get the requirements for each of the components for them can be worked together in an integrated modelling procedure. Second, corresponding to the involved disciplines, the relevant experts from different disciplines were involved and discussed together for giving the possible problems and the potential solutions for making the multi-disciplinary components to be worked together in an integrated network. The important thing was that these two perspectives of efforts were interacted and repeated along with the progress of the study, which made the expected and unexpected difficulties to be always solved under the principle of coordination and cooperation for building the integrated modelling procedure.

(III) Technical and quantified research
The process for the development of the integrated hydro-ecological modelling procedure in this study was very technical and the approaches adopted in the process were highly quantified. The general research proposal of this study was inspired by the research for the cost-effective AES design regarding biodiversity conservation by Wätzold et al. (2016), which applied very technical and quantified research methodologies. The aim of this study was to bring the same logic of it for solving another important environment problem of NPS water pollution from cropland in terms of cost-effective AES design, with highly technical tools and quantified methodologies.

Based on this, the overall structure of the integrated modelling procedure in this study was built with strict technical connection routes, as described in Chapter 4. Meanwhile, for performing the tasks in the individual components of the integrated hydro-economic modelling procedure in this study the different quantified and technical approaches were needed. For example, for the component of measure identification for the study region in Chapter 5, technical knowledge regarding agronomic crop production and engineering process for measure implementation were needed. For obtaining the quantified intermediate results of the mitigation effects and economic costs of measures, the building of the eco-hydrological model of SWAT needed highly techniques, and the cost analysis and calculation process involved lots of technical expertise in economics, as demonstrated in Chapter 6 and Chapter 7 respectively. Meanwhile, quantified data were needed in these two chapters for processing the work, with also the obtained results being presented as quantified ones. These quantified results of mitigation effects and economic costs of measures were adopted for the work of the simulation and optimization of AES, which required so highly technical knowledge in mathematical optimization that an expert in the field had to be involved.

10.2   Limitations and Future Research

(I) Data limitation

a. SWAT model simulation
During the process for setting up SWAT model and doing the simulation, the data limitation reflected in three perspectives. First, although the area of the study region was small (about 56 km^2), there were only three types of soil for simulating the study region according to the data could be best searched. This might not be the real physical situation in reality in the study region. It reduced the resolution and spatial heterogeneity for the simulation of the watershed. Besides, the changing situation of the study region along with the temporal scale in reality was not considered during the model building process in this study.

Second, when doing the calibration and validation of the SWAT model, only four years of data were available regarding the load of streamflow and sediment. Among these four years of data, two years of data were adopted for the model calibration and another two years of data were adopted for the model validation for the indices of streamflow and sediment. In general, this was relatively a small amount of data for doing the calibration and validation process to adjust the model for being as close as possible to the real situation of the study region. Besides, this study had three kinds of targeted pollutants, which are sediment, nitrogen and phosphors. When doing the calibration and validation work for the SWAT model, it should be that each of these three indices was carried out independently. However, the observed pollution load data for nitrogen and phosphorus in the corresponding monitoring station in the watershed were confidential and unavailable. Therefore, the calibration and validation process for these two kinds of targeted pollutants in this study cannot be performed.

Third, the AES design life in this study was set as a five years' period from 2018 to 2022. For simulating the mitigation effects of each measure in each heterogeneous spatial unit, the results should be the average yearly mitigation effects correspondingly for each measure based on the annual simulated results between 2018 and 2022. However, if doing this the SWAT model needed the weather input data from 2018 to 2022, which was unavailable in this study. It was dealt with the assumption that the yearly average mitigation effects of each measure during the AES design life were represented by the yearly average mitigation effects correspondingly based on the simulated results from 2013 to 2016 in the SWAT model. As the available weather data for SWAT model in this study were obtained from 2008 to 2016 for the study region.

Regarding this problem, there were two alternatives to address it generally. One was that using the SWAT model function of "Weather Generator", which could automatically generate the future several years' weather data based on the input of the last many years' weather data in the model. The problem with this was that usually at least 20 years' historical weather data were needed to simulate meteorological data in SWAT (Dixon and Earls 2012). Besides, even with enough input historical weather data, the meteorological simulation results by the "Weather Generator" of SWAT was generally not better than the results from other professional weather prediction approaches. The other alternative, therefore, was that to collect the needed weather data based on the professional weather anticipations and then input them into SWAT model for doing the corresponding simulations. However, due to the resource limitation, it was not performed in this study.

b. Economic cost calculation

Data for the calculation of the economic costs of measures were mainly obtained based on the questionnaire survey. The limitation for this was that due to the resource limitation, the number of samples for the questionnaire survey in the study region were 40 in total with eight villages. Regarding most of the data needed from the questionnaire, the small number of the samples would not have influences, as most of the data needed were related to market situations in the local area, which basically were stable among residents, such as the price of grain. However, for some data asked in the questionnaire, the number of samples might influence the accuracy more or less for representing the situation of the whole region, like the amount of time needed for finishing some kinds of labor work in the field of cropland. In addition, a few data needed for the cost calculation in this study were related to the assumed scenarios, such as the building of filter strip which farmers never applied before. Regarding this problem, some of them, like the amount of time needed for trimming the grass of filer strip was collected based on farmers' experience and expectation. Some of the data, like the price of the grass seed was collected based on the information from the online stores, which, however, would be the general situation for the whole country but not especially for the local study region.

As the aim of this study was to develop a method, the problem of data limitation for both the SWAT model simulation and the economic cost calculation had relatively less impact for the demonstration of the method development process. However, with future research, if based on the developed method one wanted to make the research results for a certain study region to be applied in practice, more sufficient and accurate data were required.

(II) Components of cost for farmers

As described in Chapter 7, the categories of economic costs for AES participation from farmers' perspective generally included the abatement costs of measures, the private transaction costs of AES participation, and the uncertainty costs for different farmers. The sum of all of these categories of costs should be the participation costs of farmers to take part in the AES program in reality.

However, in this study only the abatement costs of measures were considered, and it was assumed that the abatement costs would represent the participation costs of farmers. In the future research, it was better to consider together the other two cost categories. Meanwhile, maybe more other cost factors except these three cost categories, which also existed in reality during the process of AES program implementations, were also analyzed and considered for the calculation of AES costs for farmers.

(III) Principles for simulation of AES

As demonstrated in Chapter 8, regarding the simulation of AES, the principles for mimicking farmers' selection behavior were based on the assumption that farmers were profit-maximizing pursuers. Therefore, the only criterion to decide whether a farmer would select a measure for a spatial unit and which measure the farmer would select for that spatial unit was based on that whether in the spatial unit the measures had positive net economic benefits for famers and which measure could bring the maximum net economic benefit.

However, in practice in an AES program the factors that would influence the selection behavior of farmers regarding agri-environmental measures might be more than the consideration for the net economic benefits. In general, the possible other impact factors might include the wealth and poverty situations of farmers (like, whether a farmer feel it was worthy to earn the net economic benefits from a measure implementation), the preferences of farmers for the different measures, the perception of farmers to the cropland, and others. It would be very meaningful to consider these impact factor in the future research.

(IV) Area of spatially heterogeneous units

In this study the study region was divided as 80 SHUs, with 50 ones had land use of cropland. These 50 SHUs were the focus of this study for going through the integrated hydro-economic modelling procedure to get the cost-effective AES design. The thing was that the problem involved in this study was watershed management plan with hydrological principles, which induced that the SHUs of the study region needed to be divided by the eco-hydrological model of SWAT with the hydrological principles and resulted as the hydrological units. These spatial units are heterogeneous also in terms of the size, with some spatial units being quite big and some spatial units being quite small in comparison. Due to the big difference for the area of the spatial units, when going through the process of optimization of AES in the integrated modelling procedure, the spatial units with very big area had the possibility to be firstly selected with measures. If it was the case, the spatial units with very big area would first spend the majority of the given budget, which made that the many other spatial units left had no opportunity to be selected with measures. This indicated that the size difference of the divided SHUs might influence the cost-effectiveness of the AES for the whole study region.

In the future research, if one wanted to improve the cost-effectiveness of the resulted AES for the whole study region from this perspective, it was better to make the eco-hydrological model to try to divide the SHUs with more even size. It is workable for SWAT model. Based on the automatic division for the SHUs in SWAT, one can further manually add or delete the outlets of sub-watersheds to make the size of the final obtained SHUs to be much even, although they cannot be totally equal in size.

(V) Alternatives of cost-effectiveness
The cost-effective AES referred to two alternatives. One was that under given budget the achieved total mitigation effect of AES was maximized; the other was that with obtaining the required total mitigation effects the needed budget for the AES was minimum. In this study, only the former alternative way was applied for getting the cost-effective AES. This is because the limitation of the optimization modelling procedure in this study, which could do the simulation and optimization of AES with given budget but could not do the same with the wanted amount of total mitigation effects. However, the optimization modelling procedure for doing the simulation and optimization of AES might be improved in the future research, so that the other alternative for achieving the cost-effective AES, i.e. with the required total mitigation effect the needed budget was minimum, could also be performed.

10.3 Policy Recommendations

(I) Environment protection with cost-effective AES
Environmental problems are hard to address, especially for the NPS pollution problems. Political regulations with command and control have their limitations and shortcomings regarding environmental protection. The marketed based AES, acting as an economic instrument, is increasingly recommended by researchers as

an effective and cost-effective tool to deal with tricky environmental problems. AES is highly recommend as it attempts to solve the problem with the logic of market transaction, which makes the ESS providers to have more interest and motivation to participate in the AES programs. Meanwhile, the AES agents can make good use of the market indicators during the application of the AES program to make the results of the AES to be more cost-effective.

AES is first developed and more widely applied now in developed countries. However, in many other countries the environmental problems are much more severe, and AES is recommended for getting the positive participation of farmers. In addition, with usually the limited environmental protection funds, the design of cost-effective AES can make the limited funds to be best used. For designing cost-effective AES, one key point is to consider sufficiently the spatial heterogeneity regarding both the ecological effects and economic costs of potential agri-environmental measures. It is very important to consider the role of farmers in AES and to design the AES with the consideration of farmers' interest, if the AES agents want to pursue the cost-effectiveness. To do this, the basic thing is to obtain the asymmetric information which is clear for farmers but not for AES agents, which are the spatially heterogeneous economic costs of agri-environmental measures. The general steps for designing cost-effective AES for a certain region according to this study are recommended as below.

a. Deciding the environmental targets in the region, as well as identifying the corresponding agri-environmental measures for the region.
b. Dividing the region with a number of SHUs.
c. Assessing the quantified ecological effects and economic costs of each of the identified agri-environmental measures in each of the divided SHUs.
d. Based on the simulation of farmers' selection behavior for agri-environmental measures, to obtain the optimized set of payments of different measures in an AES that could induces the maximum total ecological effects in the whole region under given budget. This process can only be operated with the intermediate quantified results regarding ecological effects and economic costs of measures in the previous step.

(II) Suggestions for Baishahe watershed
This study aims to develop a novel method to design cost-effective AES, but not to provide a designed cost-effective AES for the study region to implement in practice. Due to the data limitation from the study region, the resulted cost-effective AES in the end of this study are not recommended as the accurate ones to be implemented in the local region. However, some recommendations based

on the general findings during the study process for the Baishahe watershed are given as below.

For mitigating the load of pollutants of sediment, nitrogen and phosphorus from cropland to waterbodies, among the identified twelve measures in this study there are different recommended measures regarding different environmental targets. First, if the environmental target is the only the pollutant of sediment, then the recommend measure is no-till in general for the Baishahe watershed. Second, when the environmental targets are either the single pollutants of nitrogen or phosphorus, or the combined targets with nitrogen and phosphorus in any ways, the situation for the recommended measures are consistent. If all twelve measures are involved the recommended ones are filter strip of 5 meters in general in the Baishahe watershed; while if the structural measures of filter strips are not accepted, the recommended measures are generally no-till and chemical fertilizer reduction by 25%. Third, when the environmental targets are the combination with different weights for the three pollutants of sediment, nitrogen and phosphorus, the recommend measures are still filter strip of 5 meters among the twelve measures. If the structural measures of filter strips are not accepted in the Baishahe watershed, then the recommended measures are also no-till and chemical fertilizer reduction by 25%. However, when the weight for sediment in the combined environmental targets is very high (more than 60%), then the measure of no-till is more recommended than chemical fertilizer reduction by 25% among the nine measures (except filter strips).

According to the report from the Environmental Protection Bureau in Xia county (Environmental Protection Bureau in Xia county (EPBX) 2012), the major problem for the Baishahe watershed is sediment caused by soil erosion, with the problem of nutrient pollutants being minor. This situation does not change in more recent years, as confirmed by the local governmental officers in Water Conservancy Bureau in Xia county (J.J. Jin, C. Guo, Water Conservancy Bureau, Xia county, personal communication, January, 2018). Based on this, the recommend measures for the Baishahe watershed are mainly no-till and filter strip of 5 meters (if structural measures are acceptable in the region).

According to the intermediate results of this study for the assessment of the abatement costs of each measure, for some SHUs in the Baishahe watershed the measure of chemical fertilizer reduction by 25% has negative results of costs. This implies that farmers actually could earn more net economic benefits if they apply the chemical fertilizer a little bit less for some patches of cropland in the Baishahe watershed. This is in fact a general situation in China, that the chemical fertilizer is excessively applied on cropland. Related research has given recommendations for the appropriate amount of chemical fertilizer for some regions in

China (Zhang et al. 2018; Xu et al. 2014; e.g. Ju et al. 2007). According to Sun et al. (2012), there are mainly two reasons for the over use of chemical fertilizer in China. One is that subsidies are provided by Chinese government for the production and distribution of synthetic fertilizers under the food self-sufficiency pressure, which lowers the cost to famers and encourage them to apply excessive fertilizer. The other is the inadequate agricultural extension services, which means that farmers have received poor training for fertilizer application and management technologies.

Based on this, it is recommended for the local government to give farmers the suggestions for the appropriate amount of chemical fertilizer application, like through the method of "soil testing and fertilizer recommendation" in Table A.4 in Appendix 1 in the Electronic Supplementary Material. Meanwhile, the price of chemical fertilizer could be adjusted according to the market situation regarding the chemical fertilizer demand from farmers as well as the amount of chemical fertilizer that is expected for farmers to apply on cropland. Besides, proper public education is needed in order to change farmers' fixed but not scientifically right awareness regarding the crop production activities, while instead to give them the instruction for the scientifically proper approaches for cropping.

Bibliography

Abbaspour, K. C.; Rouholahnejad, E.; Vaghefi, S.; Srinivasan, R.; Yang, H.; Kløve, B. (2015): A continental-scale hydrology and water quality model for Europe: Calibration and uncertainty of a high-resolution large-scale SWAT model. In *Journal of Hydrology* 524, pp. 733–752. DOI: https://doi.org/10.1016/j.jhydrol.2015.03.027.

Aghakouchak, Amir; Habib, Emad (2010): Application of a conceptual hydrologic model in teaching hydrologic processes. In *International Journal of Engineering Education* 26 (4), pp. 963–973.

Agricultural Bureau in Xia county (2015): Summary of farmland quality protection and improvement in xia county in 2015.

Ahlvik, Lassi; Ekholm, Petri; Hyytiäinen, Kari; Pitkänen, Heikki (2014): An economic–ecological model to evaluate impacts of nutrient abatement in the Baltic Sea. In *Environmental Modelling & Software* 55, pp. 164–175. DOI: https://doi.org/10.1016/j.envsoft.2014.01.027.

Aksoy, Hafzullah; Kavvas, M. Levent (2005): A review of hillslope and watershed scale erosion and sediment transport models. In *Catena* 64 (2–3), pp. 247–271. DOI: https://doi.org/10.1016/j.catena.2005.08.008.

Amundson, Ronald; Berhe, Asmeret Asefaw; Hopmans, Jan W.; Olson, Carolyn; Sztein, A. Ester; Sparks, Donald L. (2015): Soil and human security in the 21st century. In *Science* 348 (6235), p. 1261071. DOI: https://doi.org/10.1126/science.1261071.

Ando; Camm; Polasky; Solow (1998): Species distributions, land values, and efficient conservation. In *Science* 279 (5359), pp. 2126–2128. DOI: https://doi.org/10.1126/science.279.5359.2126.

Ansell, Dean; Freudenberger, David; Munro, Nicola; Gibbons, Philip (2016): The cost-effectiveness of agri-environment schemes for biodiversity conservation. A quantitative review. In *Agriculture, Ecosystems & Environment* 225, pp. 184–191. DOI: https://doi.org/10.1016/j.agee.2016.04.008.

Arabi, Mazdak; Frankenberger, Jane R.; Engel, Bernie A.; Arnold, Jeff G. (2008): Representation of agricultural conservation practices with SWAT. In *Hydrol. Process.* 22 (16), pp. 3042–3055. DOI: https://doi.org/10.1002/hyp.6890.

Arabi, Mazdak; Govindaraju, Rao S.; Hantush, Mohamed M. (2006): Cost-effective allocation of watershed management practices using a genetic algorithm. In *Water Resour. Res.* 42 (10), p. 10429. DOI: https://doi.org/10.1029/2006WR004931.

Armsworth, Paul R.; Acs, Szvetlana; Dallimer, Martin; Gaston, Kevin J.; Hanley, Nick; Wilson, Paul (2012): The cost of policy simplification in conservation incentive programs. In *Ecology letters* 15 (5), pp. 406–414. DOI: https://doi.org/10.1111/j.1461-0248.2012.017 47.x.

Arnold, J. G.; Kiniry, JR; Srinivasan, R.; Williams, JR; Haney, E. B.; Neitsch, S. L. (2013): SWAT 2012 input/output documentation. Texas Water Resources Institute. Texas Water Resources Institute.

Arnold, J. G.; Moriasi, N. d.; Gassman, P. W.; Abbaspour, K. C.; White, M. J.; Srinivasan, R. et al. (2012): SWAT: Model Use, Calibration, and Validation. In *Transactions of the ASABE* 55 (4), pp. 1491–1508. DOI: https://doi.org/10.13031/2013.42256.

Balana, Bedru B.; Jackson-Blake, Leah; Martin-Ortega, Julia; Dunn, Sarah (2015): Integrated cost-effectiveness analysis of agri-environmental measures for water quality. In *Journal of environmental management* 161, pp. 163–172. DOI: https://doi.org/10.1016/j.jenvman.2015.06.035.

Balana, Bedru Babulo; Vinten, Andy; Slee, Bill (2011): A review on cost-effectiveness analysis of agri-environmental measures related to the EU WFD. Key issues, methods, and applications. In *Ecological Economics* 70 (6), pp. 1021–1031. DOI: https://doi.org/10.1016/j.ecolecon.2010.12.020.

Bamière, Laure; Havlík, Petr; Jacquet, Florence; Lherm, Michel; Millet, Guy; Bretagnolle, Vincent (2011): Farming system modelling for agri-environmental policy design: The case of a spatially non-aggregated allocation of conservation measures. In *Ecological Economics* 70 (5), pp. 891–899. DOI: https://doi.org/10.1016/j.ecolecon.2010.12.014.

Barratt, Helen (2009): Methods of sampling from a population. Available online at https://www.healthknowledge.org.uk/public-health-textbook/research-methods/1a-epidemiology/methods-of-sampling-population, checked on 6/10/2019.

Batáry, Péter; Dicks, Lynn V.; Kleijn, David; Sutherland, William J. (2015): The role of agri-environment schemes in conservation and environmental management. In *Conservation biology : the journal of the Society for Conservation Biology* 29 (4), pp. 1006–1016. DOI: https://doi.org/10.1111/cobi.12536.

Baylis, Kathy; Peplow, Stephen; Rausser, Gordon; Simon, Leo (2008): Agri-environmental policies in the EU and United States: A comparison. In *Ecological Economics* 65 (4), pp. 753–764. DOI: https://doi.org/10.1016/j.ecolecon.2007.07.034.

Bekele, Wagayehu; Drake, Lars (2003): Soil and water conservation decision behavior of subsistence farmers in the Eastern Highlands of Ethiopia: a case study of the Hunde-Lafto area. In *Ecological Economics* 46 (3), pp. 437–451. DOI: https://doi.org/10.1016/S0921-8009(03)00166-6.

Benett, Genevieve; Carroll, Nathaniel (2014): Gaining depth: State of watershed investment 2014. Forest Trends' Ecosystem Marketplace. Washington, DC. Available online at   https://www.forest-trends.org/wp-content/uploads/2014/12/SOWI2014_Full.pdf, checked on 5/3/2019.

Bennett, Michael T. (2008): China's sloping land conversion program: Institutional innovation or business as usual? In *Ecological Economics* 65 (4), pp. 699–711. DOI: https://doi.org/10.1016/j.ecolecon.2007.09.017.

Beusen, Arthur H. W.; Bouwman, Alexander F.; van Beek, Ludovicus P. H.; Mogollón, José M.; Middelburg, Jack J. (2016): Global riverine N and P transport to ocean increased during the 20th century despite increased retention along the aquatic continuum. In *Biogeosciences* 13 (8), pp. 2441–2451. DOI: https://doi.org/10.5194/bg-13-2441-2016.

Birner, Regina; Wittmer, Heidi (2004): On the 'Efficient Boundaries of the State': The Contribution of Transaction-Costs Economics to the Analysis of Decentralization and Devolution in Natural Resource Management. In *Environ Plann C Gov Policy* 22 (5), pp. 667–685. DOI: https://doi.org/10.1068/c03101s.

Boardman, Anthony E.; Greenberg, David H.; Vining, Aidan R.; Weimer, David L. (2017): Cost-benefit analysis: concepts and practice: Cambridge University Press.

Borah, D. K.; Bera, M. (2003): Watershed-scale hydrologic and nonpoint-source pollution models: Review of mathematical bases. In *Transactions of the ASAE* 46 (6), pp. 1553–1566. DOI: https://doi.org/10.13031/2013.15644.

Borah, D. K.; Bera, M. (2004): Watershed-scale hydrologic and nonpoint-source pollution models: Review of applications. In *Transactions of the ASAE* 47 (3), pp. 789–803. DOI: https://doi.org/10.13031/2013.16110.

Boyd, Claude E. (2003): Guidelines for aquaculture effluent management at the farm-level. In *Aquaculture* 226 (1–4), pp. 101–112. DOI: https://doi.org/10.1016/S0044-8486(03)00471-X.

Brenninkmeyer, Mary Liz (1999): The ones that got away: regulating escaped fish and other pollutants from salmon fish farms. In *BC Envtl. Aff. L. Rev.* 27, pp. 75–121.

Brouwer, Roy; Hofkes, Marjan (2008): Integrated hydro-economic modelling: Approaches, key issues and future research directions. In *Ecological Economics* 66 (1), pp. 16–22. DOI: https://doi.org/10.1016/j.ecolecon.2008.02.009.

Cai, Ximing; McKinney, Daene C.; Lasdon, Leon S. (2003): Integrated Hydrologic-Agronomic-Economic Model for River Basin Management. In *Journal of Water Resources Planning and Management* 129 (1), pp. 4–17. DOI: https://doi.org/10.1061/(ASCE)0733-9496(2003)129:1(4).

Cai, Yinying; Yu, Liangliang (2018): Rural household participation in and satisfaction with compensation programs targeting farmland preservation in China. In *Journal of Cleaner Production* 205, pp. 1148–1161. DOI: https://doi.org/10.1016/j.jclepro.2018.09.011.

Campbell, Neil; D'Arcy, Brian; Frost, Alan; Novotny, Vladimir; Sansom, Anne (2005): Diffuse pollution: IWA publishing.

Cesaro, Luca; Marongiu, Sonia; Arfini, Filippo; Donati, Michele; Capelli, Maria Giacinata (2008): Cost of production. Definition and Concept. FACEPA Deliverable D1.1.2. Available online at http://facepa.slu.se/documents/Deliverable_D1-1-2_LEI.pdf, checked on 6/6/2019.

Chang, Hong-Chan; Kuo, Cheng-Chien (1994): Network reconfiguration in distribution systems using simulated annealing. In *Electric Power Systems Research* 29 (3), pp. 227–238. DOI: https://doi.org/10.1016/0378-7796(94)90018-3.

Chen, Lei; Qiu, Jiali; Wei, Guoyuan; Shen, Zhenyao (2015): A preference-based multi-objective model for the optimization of best management practices. In *Journal of Hydrology* 520, pp. 356–366. DOI: https://doi.org/10.1016/j.jhydrol.2014.11.032.

ChinaDaily (2012): China to invest billions in sandstorm controls. Available online at http:/
/www.chinadaily.com.cn/bizchina//////2012-10/08/content_15800004.htm, updated on 8/
6/2020, checked on 8/6/2020.

Ciaian, P.; PALOMA, S.; Delincé, J. (2013): Literature review on cost of production methodologies. In The First Scientific Advisory Committee Meeting. Available online at http:/
/www.fao.org/fileadmin/templates/ess/documents/meetings_and_workshops/GS_SAC_
2013/Improving_methods_for_estimating_CoP/Improving_methods_for_estimating_c
ost_of_production_in_developing_countries_Report_JRC_Lit_Review.pdf, checked on
6/6/2019.

Ciparisse, Gérard (2003): Multilingual thesaurus on land tenure: Food and Agriculture Organization of the United Nations. Available online at http://www.fao.org/3/x2038e/x20
38e0b.htm, checked on 4/28/2019.

Claassen, Roger; Cattaneo, Andrea; Johansson, Robert (2008): Cost-effective design of agri-
environmental payment programs: U.S. experience in theory and practice. In Ecological
Economics 65 (4), pp. 737–752. DOI: https://doi.org/10.1016/j.ecolecon.2007.07.032.

Commission of the European Communities (1985): Perspectives for the Common Agricultural Policy. Brussels: The Commission.

Cools, Jan; Broekx, Steven; Vandenberghe, Veronique; Sels, Hannes; Meynaerts, Erika; Ver-
caemst, Peter et al. (2011): Coupling a hydrological water quality model and an economic
optimization model to set up a cost-effective emission reduction scenario for nitrogen. In
Environmental Modelling & Software 26 (1), pp. 44–51. DOI: https://doi.org/10.1016/j.
envsoft.2010.04.017.

Dai, C.; Qin, X. S.; Tan, Q.; Guo, H. C. (2018): Optimizing best management practices for
nutrient pollution control in a lake watershed under uncertainty. In Ecological Indicators
92, pp. 288–300. DOI: https://doi.org/10.1016/j.ecolind.2017.05.016.

Daniel, Edsel B.; Camp, Janey V.; LeBoeuf, Eugene J.; Penrod, Jessica R.; Dobbins, James
P.; Abkowitz, Mark D. (2011): Watershed Modeling and its Applications: A State-of-the-
Art Review. In TOHYDJ 5 (1), pp. 26–50. DOI: https://doi.org/10.2174/187437810110
5010026.

Daniela, Ottaviani (2011): The role of PES in agriculture. Food and Agriculture Organization of United Nations, Rome, Italy. Available online at http://www.fao.org/3/i2100e/i21
00e01.pdf, checked on 5/11/2019.

D'Arcy, B.; Frost, A. (2001): The role of best management practices in alleviating water
quality problems associated with diffuse pollution. In Science of The Total Environment
265 (1–3), pp. 359–367. DOI: https://doi.org/10.1016/S0048-9697(00)00676-8.

Davis, Timothy W.; Stumpf, Richard; Bullerjahn, George S.; McKay, Robert Michael L.;
Chaffin, Justin D.; Bridgeman, Thomas B.; Winslow, Christopher (2019): Science meets
policy. A framework for determining impairment designation criteria for large waterbodies affected by cyanobacterial harmful algal blooms. In Harmful algae 81, pp. 59–64.
DOI: https://doi.org/10.1016/j.hal.2018.11.016.

Dean, Robin; Damm-Luhr, Tobia (2010): A Current review of chinese land-use law and
policy: A breakthrough in rural reform. In Pac. Rim L. & Pol'y J. 19, p. 121.

Dixon, Barnali; Earls, Julie (2012): Effects of urbanization on streamflow using SWAT with
real and simulated meteorological data. In Applied Geography 35 (1-2), pp. 174–190.
DOI: https://doi.org/10.1016/j.apgeog.2012.06.010.

Dobbs, Thomas L.; Pretty, Jules (2008): Case study of agri-environmental payments: The United Kingdom. In *Ecological Economics* 65 (4), pp. 765–775. DOI: https://doi.org/10.1016/j.ecolecon.2007.07.030.

Donald, P. F.; Evans, A. D. (2006): Habitat connectivity and matrix restoration. The wider implications of agri-environment schemes. In *Journal of Applied Ecology* 43 (2), pp. 209–218. DOI: https://doi.org/10.1111/j.1365-2664.2006.01146.x.

Dowd, B. M.; Press, D.; Los Huertos, M. (2008): Agricultural nonpoint source water pollution policy. The case of California's Central Coast. In *Agriculture, Ecosystems & Environment* 128 (3), pp. 151–161. DOI: https://doi.org/10.1016/j.agee.2008.05.014.

Drechsler, Martin; Johst, Karin; Ohl, Cornelia; Wätzold, Frank (2007a): Designing cost-effective payments for conservation measures to generate spatiotemporal habitat heterogeneity. In *Conservation biology : the journal of the Society for Conservation Biology* 21 (6), pp. 1475–1486. DOI: https://doi.org/10.1111/j.1523-1739.2007.00829.x.

Drechsler, Martin; Wätzold, Frank; Johst, Karin; Bergmann, Holger; Settele, Josef (2007b): A model-based approach for designing cost-effective compensation payments for conservation of endangered species in real landscapes. In *Biological Conservation* 140 (1–2), pp. 174–186. DOI: https://doi.org/10.1016/j.biocon.2007.08.013.

Duan, J.; Yan, Yan; Wang, D. Y.; Dong, Z. J.; Dai, F. Z. (2010): Principle analysis and method improvement on cost calculation in watershed ecological compensation (in Chinese: 流域生态补偿标准中成本核算的原理分析与方法改进). In *Acta Ecological Sinica* 30 (1), pp. 221–227 (in Chinese with English abstract).

Duffy, Michael (2014): Conservation practices for landlords. Available online at https://www.extension.iastate.edu/agdm/articles/duffy/DuffyApr14.html, checked on 6/5/2019.

Efta, James A.; Chung, Woodam (2014): Planning best management practices to reduce sediment delivery from forest roads using WEPP: Road erosion modeling and simulated annealing optimization. In *Croatian Journal of Forest Engineering: Journal for Theory and Application of Forestry Engineering* 35 (2), pp. 167–178.

Ekroos, Johan; Olsson, Ola; Rundlöf, Maj; Wätzold, Frank; Smith, Henrik G. (2014): Optimizing agri-environment schemes for biodiversity, ecosystem services or both? In *Biological Conservation* 172, pp. 65–71. DOI: https://doi.org/10.1016/j.biocon.2014.02.013.

Engel, Stefanie; Pagiola, Stefano; Wunder, Sven (2008): Designing payments for environmental services in theory and practice: An overview of the issues. In *Ecological Economics* 65 (4), pp. 663–674. DOI: https://doi.org/10.1016/j.ecolecon.2008.03.011.

Environmental Protection Bureau in Xia county (2012): Report on eco-economic zoning in Xia county.

Essenfelder, Arthur H.; Pérez-Blanco, C. Dionisio; Mayer, Alex S. (2018): Rationalizing Systems Analysis for the Evaluation of Adaptation Strategies in Complex Human-Water Systems. In *Earth's Future* 6 (9), pp. 1181–1206. DOI: https://doi.org/10.1029/2018EF000826.

European Commission (2019): Agri-environment measures. Available online at https://ec.europa.eu/info/food-farming-fisheries/sustainability-and-natural-resources/agriculture-and-environment/cap-and-environment/agri-environment-measures_en, updated on 6/12/2019, checked on 1/15/2020.

Evans, Barry M.; Lehning, David W.; Borisova, Tatiana; Corradini, Kenneth J.; Sheeder, Scott A. (2003): A generic tool for evaluating the utility of selected pollution mitigation strategies within a watershed. In, vol. 2. Proc. 7th International Conf. On Diffuse Pollution and Basin Management, Dublin, Ireland, pp. 10–17.

FAO/IIASA/ISRIC/ISSCAS/JRC (2012): Harmonized World Soil Database (version 1.2). FAO, Rome, Italy and IIASA, Laxenburg, Austria. Available online at http://webarchive. iiasa.ac.at/Research/LUC/External-World-soil-database/HTML/, updated on 2/15/2020, checked on 2/15/2020.

Fereidoon, Majid; Koch, Manfred (2016): SWAT-Model based Identification of Watershed Components in a semi-arid Region with long term Gaps in the climatological Parameters' Database. In *Proceedings of the SGEM Vienna Green*, pp. 281–288.

Ferraro, Paul J. (2008): Asymmetric information and contract design for payments for environmental services. In *Ecological Economics* 65 (4), pp. 810–821. DOI: https://doi.org/ 10.1016/j.ecolecon.2007.07.029.

Field, Richard; Tafuri, Anthony N. (2006): The Use of Best Management Practices (BMPs) in Urban Watersheds: DEStech Publications, Inc, checked on 6/6/2019.

Gassman, P. W.; Reyes, M. R.; Green, C. H.; Arnold, J. G. (2007): The Soil and Water Assessment Tool: Historical Development, Applications, and Future Research Directions. In *Transactions of the ASABE* 50 (4), pp. 1211–1250. DOI: https://doi.org/10.13031/ 2013.23637.

Geng, Runzhe; Yin, Peihong; Sharpley, Andrew N. (2019): A coupled model system to optimize the best management practices for nonpoint source pollution control. In *Journal of Cleaner Production* 220, pp. 581–592. DOI: https://doi.org/10.1016/j.jclepro.2019. 02.127.

Geospatial Data Cloud (2015): Geospatial Data Cloud site, Computer Network Information Center, Chinese Academy of Sciences. Available online at http://www.gscloud.cn/sou rces/dataset_desc/421, updated on 2/15/2020, checked on 2/15/2020.

Giri, Subhasis; Nejadhashemi, A. Pouyan; Woznicki, Sean; Zhang, Zhen (2014): Analysis of best management practice effectiveness and spatiotemporal variability based on different targeting strategies. In *Hydrol. Process.* 28 (3), pp. 431–445. DOI: https://doi.org/ 10.1002/hyp.9577.

Giri, Subhasis; Nejadhashemi, A. Pouyan; Woznicki, Sean A. (2012): Evaluation of targeting methods for implementation of best management practices in the Saginaw River Watershed. In *Journal of environmental management* 103, pp. 24–40. DOI: https://doi.org/10. 1016/j.jenvman.2012.02.033.

Hajkowicz, Stefan (2009): The evolution of Australia's natural resource management programs: Towards improved targeting and evaluation of investments. In *Land Use Policy* 26 (2), pp. 471–478. DOI: https://doi.org/10.1016/j.landusepol.2008.06.004.

Harou, Julien J.; Pulido-Velazquez, Manuel; Rosenberg, David E.; Medellín-Azuara, Josué; Lund, Jay R.; Howitt, Richard E. (2009): Hydro-economic models: Concepts, design, applications, and future prospects. In *Journal of Hydrology* 375 (3–4), pp. 627–643. DOI: https://doi.org/10.1016/j.jhydrol.2009.06.037.

Harrell, Margaret C.; Bradley, Melissa A. (2009): Data collection methods. Semi-structured interviews and focus groups. Rand National Defense Research Inst santa monica ca. Rand National Defense Research Inst santa monica ca.

He, Jun; Sikor, Thomas (2015): Notions of justice in payments for ecosystem services: Insights from China's Sloping Land Conversion Program in Yunnan Province. In *Land Use Policy* 43, pp. 207–216. DOI: https://doi.org/10.1016/j.landusepol.2014.11.011.

Hérivaux, Cécile; Orban, Philippe; Brouyère, Serge (2013): Is it worth protecting groundwater from diffuse pollution with agri-environmental schemes? A hydro-economic modeling approach. In *Journal of environmental management* 128, pp. 62–74. DOI: https://doi.org/10.1016/j.jenvman.2013.04.058.

Hox, Joop J.; Boeije, Hennie R. (2005): Data collection, primary versus secondary. In *Encyclopedia of Social Measurement* 1, pp. 593–599.

Hu, Xuetao; Chen, Jining; Zhang, Tianzhu (2002): A study on non-point source pollution models (in Chinese: 非点源污染模型研究). In *Chinese Journal of Environmental Science* 23 (3), pp. 124–228.

Huang, Chun Chang; Pang, Jiangli; Zha, Xiaochun; Su, Hongxia; Jia, Yaofeng; Zhu, Yizhi (2007): Impact of monsoonal climatic change on Holocene overbank flooding along Sushui River, middle reach of the Yellow River, China. In *Quaternary Science Reviews* 26 (17–18), pp. 2247–2264. DOI: https://doi.org/10.1016/j.quascirev.2007.06.006.

Information Center of the Ministry of Agriculture and Rural Affaires (2017): Half chemical fertilizer and half organic fertilizer have the best effect (in Chinese: 一半化肥一半有机肥 配施效果最优). Available online at http://www.agri.cn/V20/SC/jjps/201706/t20170608_5666027.htm, checked on 6/11/2019.

Iton, Ardon (2012): Cost of Production Guide. Caribbean agricultural research and development institute. Available online at http://www.cardi.org/wp-content/uploads/2011/02/Cost-of-Production-Guide-by-A-Iton-Tech-Bulletin.pdf, checked on 6/6/2019.

Jha, Manoj; Gassman, Philip W.; Secchi, Silvia; Gu, Roy; Arnold, Jeff (2004): EFFECT OF WATERSHED SUBDIVISION ON SWAT FLOW, SEDIMENT, AND NUTRIENT PREDICTIONS. In *J Am Water Resources Assoc* 40 (3), pp. 811–825. DOI: https://doi.org/10.1111/j.1752-1688.2004.tb04460.x.

Jiang, huangkun; Gao, haiying; Zhang, qi (2006): The application of BMPs for agricultural NPS pollution in China (in Chinese: 农业面源污染最佳管理措施 (BMPs) 在我国的应用) . In *Chinese journal of Agro-Environment and Development* 23 (4), pp. 64–67.

Jin, Leshan; Porras, Ina; Lopez, Alvin; Kazis, Paris (2017): Sloping Lands Conversion Programme, People's Republic of China: IIED.

Ju, X. T.; Kou, C. L.; Christie, P.; Dou, Z. X.; Zhang, F. S. (2007): Changes in the soil environment from excessive application of fertilizers and manures to two contrasting intensive cropping systems on the North China Plain. In *Environmental pollution (Barking, Essex : 1987)* 145 (2), pp. 497–506. DOI: https://doi.org/10.1016/j.envpol.2006.04.017.

Ju, Xiao-Tang; Xing, Guang-Xi; Chen, Xin-Ping; Zhang, Shao-Lin; Zhang, Li-Juan; Liu, Xue-Jun et al. (2009): Reducing environmental risk by improving N management in intensive Chinese agricultural systems. In *Proceedings of the National Academy of Sciences of the United States of America* 106 (9), pp. 3041–3046. DOI: https://doi.org/10.1073/pnas.0813417106.

Kalin, Latif; Hantush, Mohamed M. (2003): Evaluation of sediment transport models and comparative application of two watershed models. US Environmental Protection Agency,

Office of Research and Development, National Risk Management Research Laboratory. Available online at http://citeseerx.ist.psu.edu/viewdoc/download?doi=10.1.1.365. 7681&rep=rep1&type=pdf, checked on 6/6/2019.

Kirkpatrick, S.; Gelatt, C. D.; Vecchi, M. P. (1983): Optimization by simulated annealing. In *Science (New York, N.Y.)* 220 (4598), pp. 671–680. DOI: https://doi.org/10.1126/science. 220.4598.671.

Kleijn, David; Sutherland, William J. (2003): How effective are European agri-environment schemes in conserving and promoting biodiversity? In *Journal of Applied Ecology* 40 (6), pp. 947–969. DOI: https://doi.org/10.1111/j.1365-2664.2003.00868.x.

Konrad, Maria Theresia; Andersen, Hans Estrup; Thodsen, Hans; Termansen, Mette; Hasler, Berit (2014): Cost-efficient reductions in nutrient loads; identifying optimal spatially specific policy measures. In *Water Resources and Economics* 7, pp. 39–54. DOI: https://doi. org/10.1016/j.wre.2014.09.001.

Krysanova, Valentina; Arnold, Jeffrey G. (2008): Advances in ecohydrological modelling with SWAT—a review. In *Hydrological Sciences Journal* 53 (5), pp. 939–947.

Krysanova, Valentina; Srinivasan, Raghavan (2015): Assessment of climate and land use change impacts with SWAT. In *Reg Environ Change* 15 (3), pp. 431–434. DOI: https://doi.org/10.1007/s10113-014-0742-5.

Krysanova, Valentina; White, Mike (2015): Advances in water resources assessment with SWAT—an overview. In *Hydrological Sciences Journal* 22 (2), pp. 1–13. DOI: https://doi.org/10.1080/02626667.2015.1029482.

Lambert, Dayton M.; Sullivan, Patrick; Claassen, Roger; Foreman, Linda (2007): Profiles of US farm households adopting conservation-compatible practices. In *Land Use Policy* 24 (1), pp. 72–88. DOI: https://doi.org/10.1016/j.landusepol.2005.12.002.

Lapar, M. (1999): Adoption of soil conservation: the case of the Philippine uplands. In *Agricultural Economics* 21 (3), pp. 241–256. DOI: https://doi.org/10.1016/S0169-515 0(99)00028-6.

Le Coënt, Philippe (2016): Agri-environmental schemes: behavorial insights and innovative designs. Montpellier.

Le Moal, Morgane; Gascuel-Odoux, Chantal; Ménesguen, Alain; Souchon, Yves; Étrillard, Claire; Levain, Alix et al. (2019): Eutrophication. A new wine in an old bottle? In *The Science of the total environment* 651, pp. 1–11. DOI: https://doi.org/10.1016/j.scitotenv. 2018.09.139.

Leedy, Paul D.; Ormrod, Jeanne Ellis (2005): Practical research: Pearson Custom.

Lescot, Jean-Marie; Bordenave, Paul; Petit, Kevin; Leccia, Odile (2013): A spatially-distributed cost-effectiveness analysis framework for controlling water pollution. In *Environmental Modelling & Software* 41, pp. 107–122. DOI: https://doi.org/10.1016/j.envsoft. 2012.10.008.

Li, Binlin; Bicknell, K. B.; Renwick, Alan (2019): Peak phosphorus, demand trends and implications for the sustainable management of phosphorus in China. In *Resources, Conservation and Recycling* 146, pp. 316–328. DOI: https://doi.org/10.1016/j.resconrec. 2019.03.033.

Li, Chengcheng; Gao, Xubo; Wang, Yanxin (2015): Hydrogeochemistry of high-fluoride groundwater at Yuncheng Basin, northern China. In *The Science of the total environment* 508, pp. 155–165. DOI: https://doi.org/10.1016/j.scitotenv.2014.11.045.

Liu, Can; Wu, Bin (2010): Grain for Green Programme in China: Policy making and implementation. In *Briefing series* 60.

Liu, Yaoze; Engel, Bernard A.; Flanagan, Dennis C.; Gitau, Margaret W.; McMillan, Sara K.; Chaubey, Indrajeet (2017): A review on effectiveness of best management practices in improving hydrology and water quality: Needs and opportunities. In *The Science of the total environment* 601-602, pp. 580–593. DOI: https://doi.org/10.1016/j.scitotenv.2017.05.212.

Liu, Yaoze; Wang, Ruoyu; Guo, Tian; Engel, Bernard A.; Flanagan, Dennis C.; Lee, John G. et al. (2019): Evaluating efficiencies and cost-effectiveness of best management practices in improving agricultural water quality using integrated SWAT and cost evaluation tool. In *Journal of Hydrology* 577, p. 123965. DOI: https://doi.org/10.1016/j.jhydrol.2019.123965.

Lucich, Iván M.; Villena, Mauricio G.; Quinteros, Mar\'\ia José (2015): Transportation costs, agricultural expansion and tropical deforestation: Theory and evidence from Peru. In *Ciencia e investigación agraria* 42 (2), pp. 153–169.

Maringanti, Chetan; Chaubey, Indrajeet; Arabi, Mazdak; Engel, Bernard (2011): Application of a multi-objective optimization method to provide least cost alternatives for NPS pollution control. In *Environmental management* 48 (3), pp. 448–461. DOI: https://doi.org/10.1007/s00267-011-9696-2.

Maringanti, Chetan; Chaubey, Indrajeet; Popp, Jennie (2009): Development of a multiobjective optimization tool for the selection and placement of best management practices for nonpoint source pollution control. In *Water Resour. Res.* 45 (6), p. 6406. DOI: https://doi.org/10.1029/2008WR007094.

Martin-Ortega, Julia; Perni, Angel; Jackson-Blake, Leah; Balana, Bedru B.; Mckee, Annie; Dunn, Sarah et al. (2015): A transdisciplinary approach to the economic analysis of the European Water Framework Directive. In *Ecological Economics* 116, pp. 34–45. DOI: https://doi.org/10.1016/j.ecolecon.2015.03.026.

Mateo-Sagasta, J.; Ongley, E.; Hao, W.; Mei, X. (2013): Guidelines to Control Water Pollution from Agriculture in China. Decoupling water pollution from agricultural production. Rome, Italy: Food and Agriculture Organization of the United Nations. Available online at http://www.fao.org/3/i3536e/i3536e.pdf, checked on 4/17/2019.

McCann, Laura (2013): Transaction costs and environmental policy design. In *Ecological Economics* 88, pp. 253–262. DOI: https://doi.org/10.1016/j.ecolecon.2012.12.012.

Meng, Q. H.; Fu, B. J.; Yang, L. Z. (2001): Effects of land use on soil erosion and nutrient loss in the Three Gorges Reservoir Area, China. In *Soil Use and Management* 17 (4), pp. 288–291. DOI: https://doi.org/10.1111/j.1475-2743.2001.tb00040.x.

Meng, Xianyong (2016): China Meteorological Assimilation Driving Datasets for the SWAT model Version 1.1. In *Cold and Arid Regions Science Data Center at Lanzhou*. DOI: https://doi.org/10.3972/westdc.002.2016.db.

Meng, Xian-yong; Wang, Hao; Cai, Si-yu; Zhang, Xue-song; Leng, Guo-yong; Lei, Xiao-hui et al. (2017): The China Meteorological Assimilation Driving Datasets for the SWAT Model (CMADS) Application in China: A Case Study in Heihe River Basin.

Metropolis, Nicholas; Rosenbluth, Arianna W.; Rosenbluth, Marshall N.; Teller, Augusta H.; Teller, Edward (1953): Equation of State Calculations by Fast Computing Machines. In *The Journal of Chemical Physics* 21 (6), pp. 1087–1092. DOI: https://doi.org/10.1063/1.1699114.

Mettepenningen, Evy; Verspecht, Ann; van Huylenbroeck, Guido (2009): Measuring private transaction costs of European agri-environmental schemes. In *Journal of Environmental Planning and Management* 52 (5), pp. 649–667. DOI: https://doi.org/10.1080/096405609 02958206.

Mewes, Melanie; Drechsler, Martin; Johst, Karin; Sturm, Astrid; Wätzold, Frank (2015): A systematic approach for assessing spatially and temporally differentiated opportunity costs of biodiversity conservation measures in grasslands. In *Agricultural Systems* 137, pp. 76–88. DOI: https://doi.org/10.1016/j.agsy.2015.03.010.

Moriasi, Daniel N.; Pai, Naresh; Steiner, Jean L.; Gowda, Prasanna H.; Winchell, Michael; Rathjens, Hendrik et al. (2019): SWAT-LUT: A Desktop Graphical User Interface for Updating Land Use in SWAT. In *J Am Water Resources Assoc* 55 (5), pp. 1102–1115. DOI: https://doi.org/10.1111/1752-1688.12789.

Mullan, Donal (2013): Soil erosion under the impacts of future climate change. Assessing the statistical significance of future changes and the potential on-site and off-site problems. In *Catena* 109, pp. 234–246. DOI: https://doi.org/10.1016/j.catena.2013.03.007.

Naidoo, Robin; Balmford, Andrew; Ferraro, Paul J.; Polasky, Stephen; Ricketts, Taylor H.; Rouget, Mathieu (2006): Integrating economic costs into conservation planning. In *Trends in ecology & evolution* 21 (12), pp. 681–687. DOI: https://doi.org/10.1016/j.tree. 2006.10.003.

Nearing, Mark A.; Xie, Yun; Liu, Baoyuan; Ye, Yu (2017): Natural and anthropogenic rates of soil erosion. In *International Soil and Water Conservation Research* 5 (2), pp. 77–84. DOI: https://doi.org/10.1016/j.iswcr.2017.04.001.

Ning, Jicai; Gao, Zhiqiang; Lu, Qingshui (2015): Runoff simulation using a modified SWAT model with spatially continuous HRUs. In *Environ Earth Sci* 74 (7), pp. 5895–5905. DOI: https://doi.org/10.1007/s12665-015-4613-2.

NLW (2013): Overview of towns and villages in Yaofeng town, Xia county (in Chinese: 乡镇概况-运城市阳光农廉网.夏县·瑶峰镇). Available online at http://www.ycsnlw.com/xz/xz01.asp?pid=14081101, checked on 5/10/2017.

Norse, David (2005): Non-point pollution from crop production. Global, regional and national issues. In *Pedosphere* 15 (4), pp. 499–508.

Norse, David; Ju, Xiaotang (2015): Environmental costs of China's food security. In *Agriculture, Ecosystems & Environment* 209, pp. 5–14. DOI: https://doi.org/10.1016/j.agee. 2015.02.014.

NRCS (2019): Conservation Practices. Available online at https://www.nrcs.usda.gov/wps/portal/nrcs/detailfull/national/technical/cp/ncps/?cid=nrcs143_026849, checked on 6/6/2019.

Nyakatawa, E. Z.; Mays, D. A.; Tolbert, V. R.; Green, T. H.; Bingham, L. (2006): Runoff, sediment, nitrogen, and phosphorus losses from agricultural land converted to sweet-gum and switchgrass bioenergy feedstock production in north Alabama. In *Biomass and Bioenergy* 30 (7), pp. 655–664. DOI: https://doi.org/10.1016/j.biombioe.2006.01.008.

OECD (2010): Paying for Biodiversity: Enhancing the Cost-Effectiveness of Payments for Ecosystem Services: OECD Publishing.

Ongley, Edwin D.; Zhang, Xiaolan; Yu, Tao. (2010): Current status of agricultural and rural non-point source Pollution assessment in China. In *Environmental pollution (Barking, Essex : 1987)* 158 (5), pp. 1159–1168. DOI: https://doi.org/10.1016/j.envpol.2009. 10.047.

Ordóñez-Fernández, Rafaela; Repullo-Ruibérriz de Torres, Miguel Angel; Márquez-García, Javier; Moreno-García, Manuel; Carbonell-Bojollo, Rosa M. (2018): Legumes used as cover crops to reduce fertilisation problems improving soil nitrate in an organic orchard. In *European Journal of Agronomy* 95, pp. 1–13. DOI: https://doi.org/10.1016/j.eja.2018. 02.001.

Pagiola, Stefano; Platais, Gunars (2007): Payments for environmental services: from theory to practice. World Bank, Washington.

Pan, Xingliang; Xu, Linyu; Yang, Zhifeng; Yu, Bing (2017): Payments for ecosystem services in China: Policy, practice, and progress. In *Journal of Cleaner Production* 158, pp. 200–208. DOI: https://doi.org/10.1016/j.jclepro.2017.04.127.

Parajuli, P. B.; Mankin, K. R.; Barnes, P. L. (2008): Applicability of targeting vegetative filter strips to abate fecal bacteria and sediment yield using SWAT. In *Agricultural Water Management* 95 (10), pp. 1189–1200. DOI: https://doi.org/10.1016/j.agwat.2008.05.006.

Paule-Mercado, M. A.; Lee, B. Y.; Memon, S. A.; Umer, S. R.; Salim, I.; Lee, C-H (2017): Influence of land development on stormwater runoff from a mixed land use and land cover catchment. In *The Science of the total environment* 599–600, pp. 2142–2155. DOI: https://doi.org/10.1016/j.scitotenv.2017.05.081.

Permani, R. (2014): Sampling, questionnaire and interview design. Available online at https://www.adelaide.edu.au/global-food/documents/dairy-production/10-sampling-que stionnaire-interview-design-rp.pdf, updated on 6/6/2019.

Pion, Nathalie (2007): Review and assessment of agri-environment schemes on biodiversity and farming landscapes in European islands, case study of isle of Wight, Shetland and Texel. Available online at http://www.globalislands.net/userfiles/netherlands_island-farm-landscapes-report.pdf, checked on 1/15/2020.

Potter, Steve; Wang, Susan; King, Arnold (2009): Modeling Strucutral Conservation Practices. Cropland Component of the National Conservation Effects Assessment Project. Texas Agri-Life Research Center, Temple, Texas. Available online at https://www.nrcs. usda.gov/Internet/FSE_DOCUMENTS/16/nrcs143_013403.pdf, checked on 6/6/2019.

Puetz, Detlev (1993): Improving data quality in household surveys. In *Data Needs for Food Policy in Developing Countries: New Directions for Household Surveys*, pp. 173–185.

Pulido-Velazquez, Manuel; Andreu, Joaquín; Sahuquillo, Andrés; Pulido-Velazquez, David (2008): Hydro-economic river basin modelling: The application of a holistic surface–groundwater model to assess opportunity costs of water use in Spain. In *Ecological Economics* 66 (1), pp. 51–65. DOI: https://doi.org/10.1016/j.ecolecon.2007.12.016.

Qi, Honghai; Altinakar, Mustafa S. (2011): A conceptual framework of agricultural land use planning with BMP for integrated watershed management. In *Journal of environmental management* 92 (1), pp. 149–155. DOI: https://doi.org/10.1016/j.jenvman.2010.08.023.

Reckling, Moritz; Hecker, Jens-Martin; Bergkvist, Göran; Watson, Christine A.; Zander, Peter; Schläfke, Nicole et al. (2016): A cropping system assessment framework—Evaluating effects of introducing legumes into crop rotations. In *European Journal of Agronomy* 76, pp. 186–197. DOI: https://doi.org/10.1016/j.eja.2015.11.005.

Reiter, Michael A.; Saintil, Max; Yang, Ziming; Pokrajac, Dragoljub (2009): Derivation of a GIS-based watershed-scale conceptual model for the St. Jones River Delaware from habitat-scale conceptual models. In *Journal of environmental management* 90 (11), pp. 3253–3265. DOI: https://doi.org/10.1016/j.jenvman.2009.04.018.

Resources and Environmental Science Data Center (2015a): Data set on spatial distribution of first-grade rivers in China (in Chinese: 中国一级河流空间分布数据集); Provincial administrative boundary data in China in 2015a (in Chinese: 中国各省份分布图). Available online at http://www.resdc.cn/data.aspx?DATAID=221, checked on 1/20/2020.

Resources and Environmental Science Data Center (2015b): Remote sensing monitoring data for land use in China in 2015b (in Chinese: 2015年中国土地利用现状遥感监测数据). Available online at http://www.resdc.cn/data.aspx?DATAID=184, checked on 6/4/2019.

Ritter, William; Shirmohammadi, Adel (2000): Agricultural Nonpoint Source Pollution: CRC Press.

Rodriguez, Hector German; Popp, Jennie; Maringanti, Chetan; Chaubey, Indrajeet (2011): Selection and placement of best management practices used to reduce water quality degradation in Lincoln Lake watershed. In *Water Resour. Res.* 47 (1), p. 1507.

Schmitt, T. J.; Dosskey, M. G.; Hoagland, K. d. (1999): Filter Strip Performance and Processes for Different Vegetation, Widths, and Contaminants. In *Journal of Environment Quality* 28 (5), pp. 1479–1489. DOI: https://doi.org/10.2134/jeq1999.004724250028000 50013x.

Schomers, Sarah; Matzdorf, Bettina (2013): Payments for ecosystem services: A review and comparison of developing and industrialized countries. In *Ecosystem Services* 6, pp. 16–30. DOI: https://doi.org/10.1016/j.ecoser.2013.01.002.

Schöttker, Oliver; Wätzold, Frank (2018): Buy or lease land? Cost-effective conservation of an oligotrophic lake in a Natura 2000 area. In *Biodivers Conserv* 27 (6), pp. 1327–1345. DOI: https://doi.org/10.1007/s10531-017-1496-4.

Science for Environment Policy (2017): Agri-environmental schemes: how to enhance the agriculture–environment relationship. Thematic Issue 57. Issue produced for the European Commission DG Environment by the Science Communication Unit, UWE, Bristol. Available online at http://ec.europa.eu/environment/integration/research/newsalert/pdf/AES_impacts_on_agricultural_environment_57si_en.pdf, checked on 5/15/2019.

Shang, Wenxiu; Gong, Yicheng; Wang, Zhongjing; Stewardson, Michael J. (2018): Eco-compensation in China: Theory, practices and suggestions for the future. In *Journal of environmental management* 210, pp. 162–170. DOI: https://doi.org/10.1016/j.jenvman.2017.12.077.

Shao, Liqun; Chen, Haibin; Zhang, Chen; Huo, Xuexi (2017): Effects of Major Grassland Conservation Programs Implemented in Inner Mongolia since 2000 on Vegetation Restoration and Natural and Anthropogenic Disturbances to Their Success. In *Sustainability* 9 (3), p. 466. DOI: https://doi.org/10.3390/su9030466.

Shen, Z.; Bai, J.; Liao, Q.; Chen, L.; others (2014a): An overview of Chin's agricultural Non-point Source issues and the development of related research. In *Hydrology: Current Research* 5 (3).

Shen, Z.; Bai, J.; Liao, Q.; Chen, L.; others (2014b): An overview of Chin's agricultural Non-point Source issues and the development of related research. In *Hydrology: Current Research* 5 (3).

Shen, Zhenyao; Chen, Lei; Xu, Liang (2013): A topography analysis incorporated optimization method for the selection and placement of best management practices. In *PloS one* 8 (1), e54520. DOI: https://doi.org/10.1371/journal.pone.0054520.

Shen, Zhenyao; Liao, Qian; Hong, Qian; Gong, Yongwei (2012): An overview of research on agricultural non-point source pollution modelling in China. In *Separation and Purification Technology* 84, pp. 104–111. DOI: https://doi.org/10.1016/j.seppur.2011.01.018.

Shukla, Manoj (2011): Soil hydrology, land use and agriculture: measurement and modelling: Cabi.

Sidemo-Holm, William; Smith, Henrik G.; Brady, Mark V. (2018): Improving agricultural pollution abatement through result-based payment schemes. In *Land Use Policy* 77, pp. 209–219. DOI: https://doi.org/10.1016/j.landusepol.2018.05.017.

Simon, A.; Klimetz, L. (2008): Relative magnitudes and sources of sediment in benchmark watersheds of the Conservation Effects Assessment Project. In *Journal of Soil and Water Conservation* 63 (6), pp. 504–522. DOI: https://doi.org/10.2489/jswc.63.6.504.

Solomatine, D. P.; Wagener, T. (2011): Hydrological Modeling (Chapter 2.16). Available online at https://www.sciencedirect.com/science/article/pii/B9780444531995000440, checked on 6/6/2019.

Sturm, Astrid; Drechsler, Martin; Johst, Karin; Mewes, Melanie; Wätzold, Frank (2018): DSS-Ecopay—A decision support software for designing ecologically effective and cost-effective agri-environment schemes to conserve endangered grassland biodiversity. In *Agricultural Systems* 161, pp. 113–116. DOI: https://doi.org/10.1016/j.agsy.2018.01.008.

Sun, Bo; Zhang, Linxiu; Yang, Linzhang; Zhang, Fusuo; Norse, David; Zhu, Zhaoliang (2012): Agricultural non-point source pollution in China. Causes and mitigation measures. In *Ambio* 41 (4), pp. 370–379. DOI: https://doi.org/10.1007/s13280-012-0249-6.

Sun, Chen; Ren, Li (2013): Assessment of surface water resources and evapotranspiration in the Haihe River basin of China using SWAT model. In *Hydrol. Process.* 27 (8), pp. 1200–1222. DOI: https://doi.org/10.1002/hyp.9213.

Sun, Qiqi; Zhang, Chunping; Yu, Xingxiu; Li, Jianhua; Zhang, Yongkun; Gao, Yan (2013): Best management practices of agricultural non-point source pollution in China: A review (in Chinese: 中国农业面源污染最佳管理措施研究进展). In *Chinese Journal of Ecology* 32 (3), pp. 772–778 (in Chinese with English abstract).

Sun, Xinzhang; Zhou, Hailin (2008): Establishing Eco-compensation System in China: Practice, Problems and Strategies. In *China Population, Resources and Environment* 18 (5), pp. 139–143. DOI: https://doi.org/10.1016/S1872-583X(09)60019-X.

Tang, hao (2010): Study on BMPs FOR pollution control of agricultural non-point sources (in Chinese: 农业面源污染控制最佳管理措施体系研究). In *Chinese journal of Yangtze River* 41 (17), PP. 54–57 (in Chinese with English abstract).

Tarek, Amin (2010): Overview on the agri-environmental policy in Europe as a system for payment for environmental services. Available online at https://zh.scribd.com/document/32820809/Overview-on-the-Agri-Environmental-Policy-in-Europe-as-a-System-for-Payment-for-Environmental-Services, checked on 5/11/2019.

The office of Ministry of Finance; the office of Ministry of Agriculture (2016): The Guidelines for New Phase of Payments for Grassland Conservation (in Chinese: 新一轮草原生态保护补助奖励政策实施指导意见). Available online at http://www.moa.gov.cn/govpublic/CWS/201603/t20160304_5040527.htm, checked on 5/31/2019.

Tuo, Ye; Chiogna, Gabriele; Disse, Markus (2015): A Multi-Criteria Model Selection Protocol for Practical Applications to Nutrient Transport at the Catchment Scale. In *Water* 7 (12), pp. 2851–2880. DOI: https://doi.org/10.3390/w7062851.

Tuo, Ye; Duan, Zheng; Disse, Markus; Chiogna, Gabriele (2016): Evaluation of precipitation input for SWAT modeling in Alpine catchment: A case study in the Adige river basin (Italy). In *The Science of the total environment* 573, pp. 66–82. DOI: https://doi.org/10.1016/j.scitotenv.2016.08.034.

Turpin, Nadine; Bontems, Philippe; Rotillon, Gilles; Bärlund, Ilona; Kaljonen, Minna; Tattari, Sirkka et al. (2005): AgriBMPWater: systems approach to environmentally acceptable farming. In *Environmental Modelling & Software* 20 (2), pp. 187–196. DOI: https://doi.org/10.1016/j.envsoft.2003.09.004.

U.S. Department of Agriculture (2015): Summary report: 2012 National Resources Inventory, Natural Resources Conservation Service, Washington, DC, and Center for Survey Statistics and Methodology. Available online at https://www.nrcs.usda.gov/Internet/FSE_DOCUMENTS/nrcseprd396218.pdf, checked on 12/13/2019.

Uetake, Tetsuya (2013): Managing agri-environmental commons through collective action: lessons from OECD countries. In : 14th Global Conference of the International Association for the Study of the Commons, Mt. Fuji, June, pp. 3–7.

United Nations Convention to Combat Desertification (2017): Agri-environmental schemes: how to enhance the agriculture–environment relationship. Available online at https://knowledge.unccd.int/publications/agri-environmental-schemes-how-enhance-agriculture-environment-relationship, checked on 1/15/2020.

Uthes, Sandra; Matzdorf, Bettina (2013): Studies on agri-environmental measures: a survey of the literature. In *Environmental management* 51 (1), pp. 251–266. DOI: https://doi.org/10.1007/s00267-012-9959-6.

Uthes, Sandra; Matzdorf, Bettina; Müller, Klaus; Kaechele, Harald (2010a): Spatial targeting of agri-environmental measures. Cost-effectiveness and distributional consequences. In *Environmental management* 46 (3), pp. 494–509. DOI: https://doi.org/10.1007/s00267-010-9518-y.

Uthes, Sandra; Sattler, Claudia; Zander, Peter; Piorr, Annette; Matzdorf, Bettina; Damgaard, Martin et al. (2010b): Modeling a farm population to estimate on-farm compliance costs and environmental effects of a grassland extensification scheme at the regional scale. In *Agricultural Systems* 103 (5), pp. 282–293. DOI: https://doi.org/10.1016/j.agsy.2010.02.001.

van Laarhoven, Peter J. M.; Aarts, Emile H. L. (Eds.) (2010): Simulated annealing. Theory and applications. Dordrecht: Springer (Mathematics and its applications, 37).

Van Tongeren, Frank (2008): Agricultural Policy Design and Implementation. OECD Food, Agriculture and Fisheries Working Papers, No. 7, OECD Publishing, Paris. Available online at https://www.oecd-ilibrary.org/agriculture-and-food/agricultural-policy-design-and-implementation_243786286663, checked on 5/11/2019.

van Velthuizen, Harrij (2017): Harmonized World Soil Database (HWSD v 1.21). Available online at http://www.iiasa.ac.at/web/home/research/researchPrograms/water/HWSD.html, checked on 10/26/2017.

Veith, T. L.; Wolfe, M. L.; Heatwole, C. D. (2004): COST-EFFECTIVE BMP PLACEMENT: OPTIMIZATION VERSUS TARGETING. In *Transactions of the ASAE* 47 (5), pp. 1585–1594. DOI: https://doi.org/10.13031/2013.17636.

Vergamini, Daniele; White, Benedict; Viaggi, Davide (2015): Agri-Environmental Policies design in Europe, USA and Australia. is an auction more cost-effective than a self-selecting contract schedule? Available online at https://econpapers.repec.org/paper/ags aiea15/207357.htm, checked on 5/16/2019.

Waidler, David; White, Mike; Steglich, Evelyn; Wang, Susan; Williams, Jimmy; Jones, C. A.; Srinivasan, R. (2011): Conservation practice modeling guide for SWAT and APEX. Texas Water Resources Institute.

Water Conservancy Bureau in Xia county (2008): Survey and technical data of reservoir, river and drainage channels in xia county.

Wätzold, Frank; Drechsler, Martin; Johst, Karin; Mewes, Melanie; Sturm, Astrid (2016): A Novel, Spatiotemporally Explicit Ecological-economic Modeling Procedure for the Design of Cost-effective Agri-environment Schemes to Conserve Biodiversity. In *Am. J. Agr. Econ.* 98 (2), pp. 489–512. DOI: https://doi.org/10.1093/ajae/aav058.

Wätzold, Frank; Lienhoop, Nele; Drechsler, Martin; Settele, Josef (2008): Estimating optimal conservation in the context of agri-environmental schemes. In *Ecological Economics* 68 (1–2), pp. 295–305. DOI: https://doi.org/10.1016/j.ecolecon.2008.03.007.

Wätzold, Frank; Mewes, Melanie; van Apeldoorn, Rob; Varjopuro, Riku; Chmielewski, Tadeusz Jan; Veeneklaas, Frank; Kosola, Marja-Leena (2010): Cost-effectiveness of managing Natura 2000 sites: an exploratory study for Finland, Germany, the Netherlands and Poland. In *Biodivers Conserv* 19 (7), pp. 2053–2069. DOI: https://doi.org/10.1007/s10 531-010-9825-x.

Wätzold, Frank; Schwerdtner, Kathleen (2005): Why be wasteful when preserving a valuable resource? A review article on the cost-effectiveness of European biodiversity conservation policy. In *Biological Conservation* 123 (3), pp. 327–338. DOI: https://doi.org/10. 1016/j.biocon.2004.12.001.

Wildemuth, Barbara M. (2016): Applications of social research methods to questions in information and library science: ABC-CLIO.

Wilson, Jeremy D.; Evans, Andrew D.; Grice, Philip V. (2009): Bird conservation and agriculture. The bird life of farmland, grassland and heathland. Cambridge: Cambridge University Press (Ecology, biodiversity and conservation).

World Bank (2018): Real interest rate (%)—China I Data. Available online at https://data. worldbank.org/indicator/FR.INR.RINR?locations=CN, updated on 7/19/2020, checked on 12/7/2018.

Wu, J. Y.; d. Huang; Teng, W. J.; Sardo, V. I. (2010): Grass hedges to reduce overland flow and soil erosion. In *Agron. Sustain. Dev.* 30 (2), pp. 481–485. DOI: https://doi.org/10. 1051/agro/2009037.

Wu, Le; Kong, Deshuai; Jin Leshan (2019): Research on the progress of the eco-compensation mechanism in China (in Chinese: 中国生态保护补偿机制研究进展). In *Acta Ecologica Sinica* 39 (1), pp. 1–8 (in Chinese with English abstract).

Wunder, Sven (2005): Payments for environmental services: some nuts and bolts. CIFOR Occasional Paper No. 42 (Centre for International Forestry Research). Bogor, Indonesia. Available online at https://vtechworks.lib.vt.edu/bitstream/handle/10919/66932/2437_009_Infobrief.pdf?sequence=1&isAllowed=y, checked on 5/11/2019.

Wynne-Jones, Sophie (2013): Connecting payments for ecosystem services and agri-environment regulation: An analysis of the Welsh Glastir Scheme. In *Journal of Rural Studies* 31, pp. 77–86. DOI: https://doi.org/10.1016/j.jrurstud.2013.01.004.

Wyoming Department of Environmental Quality (2013): Cropland Best Management Practice Manual 2013. Available online at https://www.nrcs.usda.gov/Internet/FSE_DOC UMENTS/16/nrcs143_013403.pdf, checked on 6/6/2019.

Xia County People's Government (2020): Overview of Xiaxian County. Available online at http://www.sxxiaxian.gov.cn/sing?column=xiaxiangaikuang, updated on 6/8/2020, checked on 6/8/2020.

Xiao, Bo; Wang, Qing-hai; Wu, Ju-ying; Huang, Chuan-wei; Yu, Ding-fang (2010): Protective function of narrow grass hedges on soil and water loss on sloping croplands in Northern China. In *Agriculture, Ecosystems & Environment* 139 (4), pp. 653–664. DOI: https://doi.org/10.1016/j.agee.2010.10.011.

Xiao, Dongyang; Niu, Haipeng; Fan, Liangxin; Zhao, Suxia; Yan, Hongxuan (2019): Farmers' Satisfaction and its Influencing Factors in the Policy of Economic Compensation for Cultivated Land Protection: A Case Study in Chengdu, China. In *Sustainability* 11 (20), p. 5787. DOI: https://doi.org/10.3390/su11205787.

Xie, Hui; Chen, Lei; Shen, Zhenyao (2015): Assessment of Agricultural Best Management Practices Using Models: Current Issues and Future Perspectives. In *Water* 7 (12), pp. 1088–1108. DOI: https://doi.org/10.3390/w7031088.

Xu, Xinpeng; He, Ping; Qiu, Shaojun; Pampolino, Mirasol F.; Zhao, Shicheng; Johnston, Adrian M.; Zhou, Wei (2014): Estimating a new approach of fertilizer recommendation across small-holder farms in China. In *Field Crops Research* 163, pp. 10–17. DOI: https://doi.org/10.1016/j.fcr.2014.04.014.

Yang, Guoxiang; Best, Elly P. H. (2015): Spatial optimization of watershed management practices for nitrogen load reduction using a modeling-optimization framework. In *Journal of environmental management* 161, pp. 252–260. DOI: https://doi.org/10.1016/j.jenvman.2015.06.052.

Yang, W.; Rousseau, A. N.; Boxall, P. (2007): An integrated economic-hydrologic modeling framework for the watershed evaluation of beneficial management practices. In *Journal of Soil and Water Conservation* 62 (6), pp. 423–432.

Yang, Wanhong; Liu, Yongbo; Simmons, Jane; Oginskyy, Anatoliy; McKague, Kevin (2013a): SWAT Modelling of Agricultural BMPs and Analysis of BMP Cost Effectiveness in the Gully Creek Watershed. In *University of Guelph, Guelph, Ontario. xi.*

Yang, Wu; Liu, Wei; Viña, Andrés; Luo, Junyan; He, Guangming; Ouyang, Zhiyun et al. (2013b): Performance and prospects of payments for ecosystem services programs: evidence from China. In *Journal of environmental management* 127, pp. 86–95. DOI: https://doi.org/10.1016/j.jenvman.2013.04.019.

Yunchengshi Nonglianwang (2013): Overview of towns:Yaofeng town, xia county, Yuncheng city (in Chinese: 乡镇概况—运城市阳光农廉网—夏县, 瑶峰镇). Available online at http://www.ycsnlw.com/xz/xz01.asp?pid=14081101, checked on 9/9/2018.

Zhang, X.; Srinivasan, R.; Hao, F. (2007): Predicting Hydrologic Response to Climate Change in the Luohe River Basin Using the SWAT Model. In *Transactions of the ASABE* 50 (3), pp. 901–910. DOI: https://doi.org/10.13031/2013.23154.

Zhang, X-C John; Garbrecht, Jurgen D. (2002): PRECIPITATION RETENTION AND SOIL EROSION UNDER VARYING CLIMATE, LAND USE, AND TILLAGE AND CROPPING SYSTEMS 1. In *JAWRA Journal of the American Water Resources Association* 38 (5), pp. 1241–1253. DOI: https://doi.org/10.1111/j.1752-1688.2002.tb04345.x.

Zhang, Yitao; Wang, Hongyuan; Lei, Qiuliang; Luo, Jiafa; Lindsey, Stuart; Zhang, Jizong et al. (2018): Optimizing the nitrogen application rate for maize and wheat based on yield and environment on the Northern China Plain. In *The Science of the total environment* 618, pp. 1173–1183. DOI: https://doi.org/10.1016/j.scitotenv.2017.09.183.

Zhao, G.; Mu, X.; Wen, Z.; Wang, F.; Gao, P. (2013): Soil erosion, conservation, and Eco-environment changes in the Loess Plateau of China. In *Land Degrad. Develop.* 11 (1), pp. 499–510. DOI: https://doi.org/10.1002/ldr.2246.

Zhao, Zhanqing; Qin, Wei; Bai, Zhaohai; Ma, Lin (2019): Agricultural nitrogen and phosphorus emissions to water and their mitigation options in the Haihe Basin, China. In *Agricultural Water Management* 212, pp. 262–272. DOI: https://doi.org/10.1016/j.agwat.2018.09.002.

Zhen, L.; Li, F.; Yan, H. M.; Liu, G. H.; Liu, J. Y.; Zhang, H. Y. et al. (2014): Herders' willingness to accept versus the public sector's willingness to pay for grassland restoration in the Xilingol League of Inner Mongolia, China. In *Environ. Res. Lett.* 9 (4), p. 45003. DOI: https://doi.org/10.1088/1748-9326/9/4/045003.

Zhi, Ling; Liu, Junchang; Hua, Chun (2002): A discussion on the concept and base of replacing agriculture by afforestation (in Chinese: 退耕还林 (草) 的含义与实施基础的研究). In *World Foresry Reserach* 15 (6), pp. 69–75 (in Chinese with English abstract).

Zhou, Junyu; Gu, Baojing; Schlesinger, William H.; Ju, Xiaotang (2016): Significant accumulation of nitrate in Chinese semi-humid croplands. In *Scientific reports* 6, p. 25088. DOI: https://doi.org/10.1038/srep25088.

Zhu, Lanlan; Zhang, Chunman; Cai, Yinying (2018): Varieties of agri-environmental schemes in China: A quantitative assessment. In *Land Use Policy* 71, pp. 505–517. DOI: https://doi.org/10.1016/j.landusepol.2017.11.014.

Zhuang, Yanhua; Zhang, Liang; Du, Yun; Chen, Gang (2016): Current patterns and future perspectives of best management practices research: A bibliometric analysis. In *Journal of Soil and Water Conservation* 71 (4), 98A–104A. Available online at http://deq.wyoming.gov/media/attachments/Water%20Quality/Nonpoint%20Source/Best%20Management%20Practices/2013_wqd-wpp-Nonpoint-Source_Cropland-Best-Management-Practice-Manual.pdf.

Printed in the United States
by Baker & Taylor Publisher Services